ENERGY SCIENCE, ENGINEERING AND TECHNOLOGY

PIPELINE SAFETY AND SECURITY

Energy Science, Engineering and Technology

Additional books in this series can be found on Nova's website under the Series tab.

Additional E-books in this series can be found on Nova's website under the E-books tab.

Transportation Issues, Policies and R&D

Additional books in this series can be found on Nova's website under the Series tab.

Additional E-books in this series can be found on Nova's website under the E-books tab.

ENERGY SCIENCE, ENGINEERING AND TECHNOLOGY

PIPELINE SAFETY AND SECURITY

JONATHAN T. MILLER
AND
LISA M. GURTON
EDITORS

Nova Science Publishers, Inc.
New York

For permission to use material from this book please contact us:
Telephone 631-231-7269; Fax 631-231-8175
Web Site: http://www.novapublishers.com

Library of Congress Cataloging-in-Publication Data

Pipeline safety and security / compiled by Jonathan T. Miller and Lisa M. Gurton.
p. cm.
Includes bibliographical references and index.
ISBN 978-1-61209-147-1 (hardcover)
1. Pipelines--Safety measures--Government policy--United States. 2. Pipelines--Security measures--Government policy--United States. 3. Pipelines--Accidents--United States--Prevention. 4. Pipelines--Environmental aspects. I. Miller, Jonathan Tristram, 1964- II. Gurton, Lisa M.
TJ930.P5686 2011
388.5--dc22

2010047640

Published by Nova Science Publishers, Inc. † New York

Contents

PREFACE

This new book explores how the various elements of U.S. pipeline safety and security activity fit together in the nation's overall strategy to protect transportation infrastructure.

Chapter 1- Nearly half a million miles of oil and natural gas transmission pipeline crisscross the United States. While an efficient and fundamentally safe means of transport, many pipelines carry hazardous materials with the potential to cause public injury and environmental damage. The nation's pipeline networks are also widespread, running alternately through remote and densely populated regions; consequently, these systems are vulnerable to accidents and terrorist attack.

Chapter 2- The United States depends on a vast network of pipelines to transport energy. GAO was asked to review the Transportation Security Administration's (TSA) efforts to help ensure pipeline security. This report addresses the extent to which TSA's Pipeline Security Division (PSD) has (1) assessed risk and prioritized efforts to help strengthen pipeline security, (2) implemented agency guidance and requirements of the Implementing Recommendations of the 9/11 Commission Act of 2007 (9/11 Commission Act) regarding pipeline security, and (3) measured its performance in strengthening pipeline security. GAO reviewed PSD's risk assessment process and performance measures and observed 14 PSD reviews and inspections scheduled during the period of GAO's review. Although these observations are not generalizable, they provided GAO an understanding of how PSD conducts reviews and inspections.

Chapter 3 – This is a statement of the Subcommittee on Railroads, Pipelines, and Hazardous Materials Staff, before the members of the Subcommittee on Railroads, Pipelines, and Hazardous Materials.

Chapter 4 – This is a statement of Cynthia L. Quarterman, Administrator, Pipeline and Hazardous Materials Safety Administration, before the Subcommittee on Transportation, and Hazardous Materials, Committee on Transportation and Infrastructure, United States House of Representatives.

Chapter 5 – These This is a statement of the Subcommittee on Railroads, Pipelines, and Hazardous Materials Staff, before the members of the Subcommittee on Railroads, Pipelines, and Hazardous Materials.

Chapter 6 – This is a statement of Cynthia L. Quarterman, Administrator, Pipeline and Hazardous Materials Safety Administration, before the Subcommittee on Railroads, Pipelines and Hazardous Materials.

Chapter 7 – This is a statement of Deborah A. P. Hersman, Chairman, national Transportation Safety Board, before the Subcommitttee on Railroads, Pipelines, and Hazardous Materials.

Chapter 8- This report presents the results of our review on multi-agency actions to implement the pipeline security annex to the Memorandum of Understanding (MOU) between the Department of Transportation (DOT) and Department of Homeland Security (DHS). The Pipeline Inspection, Protection, Enforcement, and Safety Act of 2006[1] requires the Office of Inspector General to assess actions taken to implement the annex.

In: Pipeline Safety and Security
Editor: Jonathan T. Miller and Lisa M. Gurton
ISBN: 978-1-61209-147-1

Chapter 1

PIPELINE SAFETY AND SECURITY: FEDERAL PROGRAMS

Paul W. Parfomak

SUMMARY

Nearly half a million miles of oil and natural gas transmission pipeline crisscross the United States. While an efficient and fundamentally safe means of transport, many pipelines carry hazardous materials with the potential to cause public injury and environmental damage. The nation's pipeline networks are also widespread, running alternately through remote and densely populated regions; consequently, these systems are vulnerable to accidents and terrorist attack.

The 109th Congress passed the Pipeline Safety Improvement Act of 2006 (P.L. 109-468) to improve pipeline safety and security practices. The 110th Congress passed the Implementing Recommendations of the 9/11 Commission Act of 2007 (P.L. 110-53), which mandated pipeline security inspections and potential enforcement (§ 1557) and required federal plans for critical pipeline security and incident recovery (§ 1558). The 111th Congress is overseeing the implementation of these acts and considering new legislation related to the nation's pipeline network. Recent legislative proposals include the Clean, Affordable, and Reliable Energy Act of 2009 (S. 1333), which would change natural gas pipeline integrity assessment intervals (§ 401); the Transportation Security Administration Authorization Act (H.R. 2220), which would mandate a new federal pipeline security study (§ 406); and the Hazardous Material Transportation Safety Act of 2009 (H.R. 4106), which seeks to improve the collection and use of hazardous material transportation incident data (§ 203) and increase staffing at the Pipeline and Hazardous Material Safety Administration (§304).

The Pipeline and Hazardous Materials Safety Administration (PHMSA), within the Department of Transportation (DOT), is the lead federal regulator of pipeline safety. PHMSA uses a variety of strategies to promote compliance with its safety regulations, including inspections, investigation of safety incidents, and maintaining a dialogue with pipeline operators. The agency clarifies its regulatory expectations through a range of communications and relies upon a range of enforcement actions to ensure that pipeline operators correct safety violations and take preventive measures to preclude future problems. The Transportation Security Administration (TSA), within the Department of Homeland Security (DHS), is the lead federal agency for security in all modes of transportation—including pipelines. The agency oversees industry's identification and protection of pipelines by developing security standards; implementing measures to mitigate security risk; building stakeholder relations; and monitoring compliance with security standards, requirements, and regulation. While PHMSA and TSA have distinct missions, pipeline safety and security are intertwined.

Although pipeline impacts on the environment remain a concern of some public interest groups, both federal government and industry representatives suggest that federal pipeline programs have been on the right track. As oversight of the federal role in pipeline safety and security continues, Congress may focus on the effectiveness of state pipeline damage prevention programs, the promulgation of low-stress pipeline regulations, federal pipeline safety enforcement, and the relationship between DHS and the DOT with respect to pipeline security, among other provisions in federal pipeline safety regulation. In addition to these specific issues, Congress may wish to assess how the various elements of U.S. pipeline safety and security activity fit together in the nation's overall strategy to protect transportation infrastructure.

INTRODUCTION[1]

Nearly half a million miles of oil and natural gas transmission pipeline crisscross the United States.[2] These pipelines are integral to U.S. energy supply and have vital links to other critical infrastructure, such as power plants, airports, and military bases. While an efficient and fundamentally safe means of transport, many pipelines carry volatile, flammable, or toxic materials with the potential to cause public injury and environmental damage. The nation's pipeline networks are also widespread, running alternately through remote and densely populated regions; consequently, these systems are vulnerable to accidents and terrorist attack. The 2006 partial shutdown of the Prudhoe Bay, Alaska oil field—the largest in the United States—due to pipeline safety problems demonstrated this vulnerability.

The 109th Congress passed the Pipeline Safety Improvement Act of 2006 (P.L. 109-468) to improve pipeline safety and security practices. The 110th Congress passed the Implementing Recommendations of the 9/11 Commission Act of 2007 (P.L. 110-53), which mandated pipeline security inspections and potential enforcement (§ 1557) and required federal plans for critical pipeline security and incident recovery (§ 1558). The 111th Congress is overseeing the implementation of these acts and considering new legislation related to the nation's pipeline network. Recent legislative proposals include the Clean, Affordable, and Reliable Energy Act of 2009 (S. 1333), which would change natural gas pipeline integrity assessment intervals (§ 401); the Transportation Security Administration Authorization Act (H.R. 2220), which would mandate a new federal pipeline security study (§ 406); and the Hazardous Material Transportation Safety Act of 2009 (H.R. 4106), which seeks to improve the collection and use of hazardous material transportation incident data (§ 203) and would increase staffing at the Pipeline and Hazardous Material Safety Administration (§304).

Pipeline Industry Characteristics

Roughly 170,000 miles of oil pipeline in the United States carry over 75% of the nation's crude oil and around 60% of its refined petroleum products.[3] There are nearly 200 *inter*state oil pipelines, which account for roughly 80% of total pipeline mileage and transported volume.[4] The U.S. natural gas pipeline network consists of around 217,000 miles of *inter*state transmission, and 89,000 miles of *intra*state transmission.[5] It also contains some 20,000 miles of field and gathering pipeline, which connect gas extraction wells to processing facilities.[6] Around 120 systems make up the *inter*state network; another 90 or so systems operate strictly within individual states.[7] These *inter*state and *intra*state gas transmission pipelines feed around 1.1 million miles of regional lines in some 1,300 local distribution networks.[8] Natural gas pipelines also connect to 113 liquefied natural gas (LNG) storage sites, which augment pipeline gas supplies during peak demand periods.[9]

Pipeline Safety Record

Taken as a whole, releases from pipelines cause few annual fatalities compared to other product transportation modes. Oil pipelines reported an average of 2.6 deaths per year from 2004 through 2008; natural gas transmission pipelines reported an average of 1.0 death per year during the same period.[10] Accidental pipeline releases result from a variety of causes, including third-party excavation, corrosion, mechanical failure, control system failure, and operator error. Natural forces, such as floods and earthquakes, can also damage pipelines. According to the Department of Transportation (DOT), there were 103 oil pipeline accidents and 63 natural gas transmission pipeline accidents in 2008.[11] Although pipeline releases have caused relatively few fatalities in absolute numbers, a single pipeline accident can be catastrophic. For example, a 1999 gasoline pipeline explosion in Bellingham, Washington, killed two children and an 18-year-old man, and caused $45 million in damage to a city water plant and other property. In 2000, a natural gas pipeline explosion near Carlsbad, New Mexico, killed 12 campers, including four children.[12] In 2006, damaged pipelines on the North Slope of Alaska leaked over 200,000 gallons of crude oil in an environmentally sensitive area. Such accidents have generated substantial scrutiny of pipeline regulation and increased state and community activity related to pipeline safety.[13]

Pipeline Security Risks

Pipelines are vulnerable to vandalism and terrorist attack with firearms, with explosives, or by other physical means. Some pipelines may also be vulnerable to "cyber-attacks" on computer control systems or attacks on electricity grids or telecommunications networks.[14] Oil and gas pipelines have been a target of terrorists globally, including within the United States. In Colombia, for example, rebels have bombed the Caño Limón oil pipeline and other pipelines over 950 times since 1993.[15] In 1996, London police foiled a plot by the Irish Republican Army to bomb gas pipelines and other utilities across the city.[16] Militants in Nigeria have repeatedly attacked pipelines and related facilities, including the simultaneous bombing of three oil pipelines in May, 2007.[17] A Mexican rebel group similarly detonated bombs along Mexican oil and natural gas pipelines in July and September, 2007.[18] In June, 2007, the U.S. Department of Justice arrested members of a terrorist group planning to attack jet fuel pipelines and storage tanks at the John F. Kennedy (JFK) International Airport in New

York.[19] Natural gas pipelines in British Columbia, Canada, were bombed six times between October 2008 and July 2009 by unknown perpetrators.[20]

Since September 11, 2001, federal warnings about Al Qaeda have mentioned pipelines specifically as potential terror targets in the United States.[21] One U.S. pipeline of particular concern and with a history of terrorist and vandal activity is the Trans Alaska Pipeline System (TAPS), which transports crude oil from Alaska's North Slope oil fields to the marine terminal in Valdez. TAPS runs some 800 miles and delivers nearly 17% of United States domestic oil production.[22] In 1999, Vancouver police arrested a man planning to blow up TAPS for personal profit in oil futures.[23] In 2001, a vandal's attack on TAPS with a high-powered rifle forced a two-day shutdown and caused extensive economic and ecological damage.[24] In January 2006, federal authorities acknowledged the discovery of a detailed posting on a website purportedly linked to Al Qaeda that reportedly encouraged attacks on U.S. pipelines, especially TAPS, using weapons or hidden explosives.[25] In November 2007 a U.S. citizen was convicted of trying to conspire with Al Qaeda to attack TAPS and a major natural gas pipeline in the eastern United States.[26] To date, there have been no known Al Qaeda attacks on TAPS or other U.S. pipelines, but such attacks remain a possibility.

Office of Pipeline Safety

The Natural Gas Pipeline Safety Act of 1968 (P.L. 90-481) and the Hazardous Liquid Pipeline Act of 1979 (P.L. 96-129) are two of the key early acts establishing the federal role in pipeline safety. Under both statutes, the Transportation Secretary is given primary authority to regulate key aspects of interstate pipeline safety: design, construction, operation and maintenance, and spill response planning. Pipeline safety regulations are covered in Title 49 of the *Code of Federal Regulations*.[27] The DOT administers pipeline regulations through the Office of Pipeline Safety (OPS) within the Pipelines and Hazardous Materials Safety Administration (PHMSA). The OPS has approximately 200 full-time equivalent staff, including inspectors, based in Washington, D.C., Atlanta, Kansas City, Houston, and Denver.[28] In addition to its own staff, PHMSA's enabling legislation allows the agency to delegate authority to *intra*state pipeline safety offices, and allows state offices to act as "agents" administering *inter*state pipeline safety programs (excluding enforcement) for those sections of *inter*state pipelines within their boundaries.[29] Over 400 state pipeline safety inspectors are available in 2010.

PHMSA's pipeline safety program is funded primarily by user fees assessed on a per-mile basis on each regulated pipeline operator (49 U.S.C. § 60107). P.L. 109-468 authorizes annual pipeline safety program expenditures (§ 18) of $79.0 million in FY2007, $86.2 million in FY2008, $91.5 million in FY2009, and $96.5 million in FY2010. The President's FY2010 budget request included $105.2 million for pipeline safety.[30] The FY2011 budget requests $111.1 million.[31]

PHMSA uses a variety of strategies to promote compliance with its safety standards. The agency conducts physical inspections of facilities and construction projects; conducts programmatic inspections of management systems, procedures, and processes; investigates safety incidents, and maintains a dialogue with pipeline operators. The agency clarifies its regulatory expectations through published protocols and regulatory orders, guidance manuals, and public meetings. PHMSA relies upon a range of enforcement actions, including administrative actions and civil penalties, to ensure that operators correct safety violations and take measures to preclude future safety problems. From 2005 through 2009, PHMSA initiated over 1,300 enforcement actions against pipeline operators. Civil penalties proposed by PHMSA for safety violations during this period totaled approximately $27.4 million.[32] PHMSA also conducts accident investigations and systemwide reviews focusing on high-risk operational or procedural problems and areas of the pipeline near sensitive environmental areas, high-density populations, or navigable waters.

Since 1997, PHMSA has increasingly encouraged industry's implementation of "integrity management" programs on pipeline segments near "high consequence" areas. Integrity management provides for continual evaluation of pipeline condition; assessment of risks to the pipeline; inspection or testing; data analysis; and followup repair, as well as preventive or mitigative actions. High-consequence areas include population centers, commercially navigable waters, and environmentally sensitive areas, such as drinking water supplies or ecological reserves. The integrity management approach directs priority resources to locations of highest consequence rather than applying uniform treatment to the entire pipeline network.[33] PHMSA made integrity management programs mandatory for most operators with 500 or more miles of regulated oil pipeline as of March 31, 2001 (49 C.F.R. § 195).

Pipeline Safety Improvement Act of 2002

On December 12, 2002, President Bush signed into law the Pipeline Safety Improvement Act of 2002 (P.L. 107-355). The act strengthened federal pipeline safety programs, state oversight of pipeline operators, and public education regarding pipeline safety.[34] Among other provisions, P.L. 107-355 required operators of regulated gas pipelines in high-consequence areas to conduct risk analysis and implement integrity management programs similar to those required for oil pipelines.[35] The act authorized the DOT to order safety actions for pipelines with potential safety problems (§ 7) and increased violation penalties (§ 8). The act streamlined the permitting process for emergency pipeline restoration by establishing an interagency committee, including the DOT, the Environmental Protection Agency, the Bureau of Land Management, the Federal Energy Regulatory Commission, and other agencies, to ensure coordinated review and permitting of pipeline repairs (§ 16). The act required DOT to study ways to limit pipeline safety risks from population encroachment and ways to preserve environmental resources in pipeline rights-of-way (§ 11). P.L. 107-355 also included provisions for public education, grants for community pipeline safety studies, "whistle blower" and other employee protection, employee qualification programs, and mapping data submission.

OPS Pipeline Security Activities

Presidential Decision Directive 63 (PDD-63), issued during the Clinton administration, assigned lead responsibility for pipeline security to the DOT.[36] At the time, these responsibilities fell to the OPS, at that time a part of the DOT's Research and Special Programs Administration (RSPA), since the agency was already addressing some elements of pipeline security in its role as safety regulator. In 2002, the OPS conducted a vulnerability assessment to identify critical pipeline facilities and worked with industry groups and state pipeline safety organizations "to assess the industry's readiness to prepare for, withstand and respond to a terrorist attack.... "[37] Together with the Department of Energy and state pipeline agencies, the OPS promoted the development of consensus standards for security measures tiered to correspond with the five levels of threat warnings issued by the Office of Homeland Security.[38] The OPS also developed protocols for inspections of critical facilities to ensure that operators implemented appropriate security practices. To convey emergency information and warnings, the OPS established a variety of communication links to key staff at the most critical pipeline facilities throughout the country. The OPS also began identifying near-term technology to enhance deterrence, detection, response, and recovery, and began seeking to advance public and private sector planning for response and recovery.[39]

On September 5, 2002, the OPS circulated formal guidance developed in cooperation with the pipeline industry associations defining the agency's security program recommendations and implementation expectations. This guidance recommended that operators identify critical facilities, develop security plans consistent with prior trade association security guidance, implement these plans, and review them annually.[40] While the guidance was voluntary, the OPS expected compliance and informed operators of its intent to begin reviewing security programs within 12 months, potentially as part of more comprehensive safety inspections.[41]

In November 2004, the President signed the Norman Y. Mineta Research and Special Programs Improvement Act (P.L. 108-426), which eliminated RSPA and placed the Office of Pipeline Safety under the new Pipeline and Hazardous Material Safety Administration. This administrative restructuring did not significantly affect the authorities or activities of the OPS.

Transportation Security Administration

In November 2001, President Bush signed the Aviation and Transportation Security Act (P.L. 107-71) establishing the Transportation Security Administration (TSA) within the DOT. According to TSA, the act placed the DOT's pipeline security authority (under PDD-63) within TSA. The act specified for TSA a range of duties and powers related to general transportation security, such as intelligence management, threat assessment, mitigation, security measure oversight and enforcement, among others. On November 25, 2002, President Bush signed the Homeland Security Act of 2002 (P.L. 107-296) creating the Department of Homeland Security (DHS). Among other provisions, the act transferred to DHS the Transportation Security Administration from the DOT (§ 403). On December 17, 2003, President Bush issued Homeland Security Presidential Directive 7 (HSPD-7), clarifying executive agency responsibilities for identifying, prioritizing, and protecting critical infrastructure.[42] HSPD-7 maintains DHS as the lead agency for pipeline security (par. 15), and instructs the DOT to "collaborate in regulating the transportation of hazardous materials by all modes (including pipelines)" (par. 22h). The order requires that DHS and other federal agencies collaborate with "appropriate private sector entities" in sharing information and protecting critical infrastructure (par. 25). TSA joined both the Energy Government Coordinating Council and the Transportation Government Coordinating Council under provisions in HSPD-7. The missions of the councils are to work with their industry counterparts to coordinate critical infrastructure protection programs in the energy and transportation sectors, respectively, and to facilitate the sharing of security information.

TSA Pipeline Security Plan

HSPD-7 also required DHS to develop a national plan for critical infrastructure and key resources protection (par. 27), which the agency issued in 2006 as the *National Infrastructure Protection Plan* (NIPP). The NIPP, in turn, required each critical infrastructure sector to develop a Sector Specific Plan (SSP) that describes strategies to protect its critical infrastructure, outlines a coordinated approach to strengthen its security efforts, and determines appropriate funding for these activities. Executive Order 13416 further required the transportation sector SSP to prepare annexes for each mode of surface transportation with the following information:

- identification of existing security guidelines, requirements, and gaps,
- description of how the SSP plan will be implemented for each mode,
- respective roles of government entities and the private sector,
- processes for review of information sharing mechanisms, and
- processes for assessing security guideline compliance and revision.[43]

In accordance with the above requirements the TSA issued its *Transportation Systems Sector Specific Plan* and *Pipeline Modal Annex* in May, 2007.

TSA Pipeline Security Activities

Pipeline security activities at TSA are led by the Pipeline Security Division (PSD) within the agency's Office of Transportation Sector Network Management.[44] According to the agency's *Pipeline Modal Annex* (PMA), TSA has been engaged in a number of specific pipeline security initiatives since 2003 as summarized in **Table 1**.

In 2003, TSA initiated its Corporate Security Review (CSR) program, wherein the agency visits the largest pipeline and natural gas distribution operators to review their security plans and inspect their facilities. During the reviews, TSA evaluates whether each company is following the intent of the OPS security guidance, and seeks to collect the list of assets each company had identified meeting the criteria established for critical facilities. In 2004, the DOT reported that the plans reviewed to date (approximately 25) had been "judged responsive to the OPS guidance."[45] As of February 2010, TSA had completed CSR's covering all of the largest 100 pipeline systems (84% of total U.S. energy pipeline throughput) and had completed revisits of 41 systems determined to be at highest risk. The agency plans to conduct 12 additional CSR's in 2010.[46] According to TSA, recent CSR results indicate that the majority of U.S. pipeline systems "continue to do a good job in regards to pipeline security" although there are areas in which pipeline security can be improved.[47] Past CSR reviews have identified inadequacies in some company security programs such as not updating security plans, lack of management support, poor employee involvement, inadequate threat intelligence, and employee apathy or error.[48]

Table 1. TSA Pipeline Security Initiatives

Initiative	Description	Participants[a]
Pipeline Policy and Planning	Coordination, development, implementation, and monitoring of pipeline security plans	TSA, DHS, DOT, DOE
Sector Coordinating Councils and Joint Sector Committee	Government partners coordinate interagency and cross-jurisdictional implementation of critical infrastructure security	TSA, DOE, Other agencies, Industry
Corporate Security Reviews (CSR)	On-site reviews of pipeline operator security	TSA, Industry
Pipeline System Risk Tool	Statistical tool used for relative risk ranking and prioritizing CSR findings	TSA, Industry
Pipeline Cross-Border Vulnerability Assessment	U.S. and Canadian security assessment and planning for critical cross-border pipeline	TSA, Canada
Regional Gas Pipeline Studies	Regional supply studies for key natural gas markets	TSA, DOE, INGAA, GTI, NETL, Industry
Cyber Attack Awareness	Training/presentations on Supervisory Control and Data Acquisition (SCADA) system vulnerabilities	TSA, GTI
Landscape Depiction and Analysis Tool	Incorporates depiction of the pipeline domain with risk analysis components	TSA
International Pipeline Security Forums	International forums for U.S. and Canadian governments and pipeline industry officials convened annually	TSA, Canada, Other agencies, Industry
"G8" Multinational Security Assessment and Planning	Multinational-sharing of pipeline threat assessment methods, advisory levels, effective practices, and vulnerability information; also develops a G8-based contingency planning guidance document	TSA, DHS, State Dept., G8 Nations
Pipeline Security Drills	Facilitation of pipeline security drills and exercises	TSA, Industry
Security Awareness Training	Informational compact discs about pipeline security issues and improvised explosive devices	TSA
Stakeholder Conference Calls	Periodic information-sharing conference calls between key pipeline security stakeholders	TSA, Other agencies, Industry
Pipeline Blast Mitigation Studies	Explosives tests on various pipe configurations to determine resiliency characteristics	TSA, DOD, Other agencies
Virtual Library Pipeline Site	Development of TSA information-sharing Web portal	TSA

Sources: Transportation Security Administration, Pipeline Modal Annex, June 2007, pp. 10-11, http://www.dhs.gov/xlibrary/assets/Transportation_Pipeline_Modal_Annex_5_21_07.pdf; Transportation Security Administration, Personal communication, February 2, 2010.

a. Key: DHS = Dept. Of Homeland Security, DOE = Dept. of Energy, G8 = Group of Eight (U.S., U.K., Canada, France, Germany, Italy, Japan, and Russia),GTI = Gas Technology Institute, INGAA = Interstate Natural Gas Association of America, NETL = National Energy Technology Laboratory, TSA = Transportation Security Administration.

In addition to the initiatives in **Table 1**, TSA has worked to establish qualifications for personnel seeking unrestricted access to critical pipeline assets and has developed its own inventory of critical pipeline infrastructure.[49] The agency has also addressed legal issues regarding recovery from terrorist attacks, such as FBI control of crime scenes and eminent domain in pipeline restoration. In October 2005, TSA issued an overview of recommended security practices for pipeline operators "for informational purposes only ... not intended to replace security measures already implemented by individual companies."[50] The agency released revised guidance on security best practices at the end of 2006, and is currently reviewing an updated version for possible release in 2010.[51] The guidelines include a section on cybersecurity developed with the assistance of the Applied Physics Laboratory of John Hopkins University as well as other government and industry stakeholders.[52]

The mission of TSA's Pipeline Security Division (PSD) currently includes developing security standards; implementing measures to mitigate security risk; building and maintaining stakeholder relations, coordination, education and outreach; and monitoring compliance with security standards, requirements, and regulations. The President's FY2011 budget request for DHS does not include a separate line item for TSA's pipeline security activities. The budget request does include a $137.6 million line item for "Surface Transportation Security," which encompasses security activities in non-aviation transportation modes, including pipelines.[53] The PSD has traditionally received from the agency's general operational budget an allocation for routine operations such as regulation development, travel, and outreach. According to the PSD, the budget funds 13 full-time equivalent staff within the office. These staff will conduct pipeline security inspections, maintain TSA's asset database, support TSA's multi-modal risk models, develop new security standards, and issue regulations, as required.[54]

In January, 2007 testimony before Congress, the TSA Administrator stated that the agency intended to conduct a pipeline infrastructure study to identify the "highest risk" pipeline assets, building upon such a list developed through the CSR program. He also stated that the agency would use its ongoing security review process to determine the future implementation of baseline risk standards against which to set measurable pipeline risk reduction targets.[55] Provisions in the Implementing Recommendations of the 9/11 Commission Act of 2007 (P.L. 110-53) require TSA, in consultation with PHMSA, to develop a plan for the federal government to provide increased security support to the "most critical" pipelines at high or severe security alert levels and when there is specific security threat information relating to such pipeline infrastructure (§ 1558(a)(1)). The act also requires a recovery protocol plan in the event of an incident affecting the interstate and intrastate pipeline system (§ 1558(a)(2)). According to TSA, a draft plan has been completed and is currently under review in the TSA/DHS clearance process.[56]

In December 2008, the Senate Committee on Commerce, Science, and Transportation requested a study by the Government Accountability Office (GAO) examining TSA's efforts to ensure pipeline security. Among other issues, the study will examine TSA's use of risk assessment and risk information in securing pipelines, actions the agency has taken to improve pipeline security under guidance in the 9/11 Commission Act of 2007 (P.L. 110-53), and the agency's efforts to measure such security improvement efforts. GAO anticipates completing its study in June 2010.[57]

Security Incident Investigations

In addition to the above pipeline security initiatives, the TSA Pipeline Security Division has performed a limited number of vulnerability assessments and has supported investigations for specific companies and assets where intelligence information has suggested potential terrorist activity. The PSD, along with the PHMSA, was involved in the investigation of an August, 2006 security breach at an LNG peak-shaving plant in Lynn, MA.[58] Although not a terrorist incident, the security breach involved the penetration of intruders through several security barriers and alert systems, permitting them to access the main LNG storage tank at the facility. The PSD also became aware of the JFK airport terrorist plot in its early stages and supported the Federal Bureau of Investigation's associated investigation. The PSD engaged the private sector in helping to assess potential targets and determine potential consequences. The PSD worked with the pipeline company to keep it informed about the plot, discuss its security practices, and review its emergency response plans.[59]

Federal Energy Regulatory Commission

One area related to pipeline safety and security not under either PHMSA's or TSA's primary jurisdiction is the siting approval of new gas pipelines, which is the responsibility of the Federal Energy Regulatory Commission (FERC). Companies building interstate natural gas pipelines must first obtain from FERC certificates of public convenience and necessity. (FERC does not oversee oil pipeline construction.) FERC must also approve the abandonment of gas facility use and services. These approvals may include safety and security provisions with respect to pipeline routing, safety standards and other factors.[60] As a practical matter, however, FERC has traditionally left these considerations to the other agencies.[61]

On September 14, 2001, the Federal Energy Regulatory Commission (FERC) notified FERC-regulated companies that it would "approve applications proposing the recovery of prudently incurred costs necessary to further safeguard the nation's energy systems and infrastructure" in response to the terror attacks of 9/11. FERC also committed to "expedite the processing on a priority basis of any application that would specifically recover such costs from wholesale customers." Companies could propose a surcharge over currently existing rates or some other cost recovery method.[62] In FY2005, the commission processed security cost recovery requests from 14 oil pipelines and 3 natural gas pipelines.[63] The FERC's FY2006 annual report states that "the Commission continues to give the highest priority to deciding any requests made for the recovery of extraordinary expenditures to safeguard the reliability and security of the Nation's energy transportation systems and energy supply infrastructure."[64] The FY2007 and FY2008 annual reports do not mention pipeline security.

In February 2003, FERC handed down a new rule (RM02-4-000) to protect critical energy infrastructure information (CEII). The rule defines CEII as information that "must relate to critical infrastructure, be potentially useful to terrorists, and be exempt from disclosure under the Freedom of Information Act." According to the rule, critical infrastructure is "existing and proposed systems and assets, whether physical or virtual, the incapacity or destruction of which would negatively affect security, economic security, public health or safety, or any combination of those matters." CEII excludes "information that identifies the location of infrastructure." The rule also establishes procedures for the public to request and obtain such critical information, and applies both to proposed and existing infrastructure.[65]

On May 14, 2003, FERC handed down new rules (RM03-4) facilitating the restoration of pipelines after a terrorist attack. The rules allow owners of a damaged pipeline to use blanket certificate authority to immediately start rebuilding, regardless of project cost, even outside existing rights-of-way. Pipeline owners would still need to notify landowners and comply with environmental laws. Prior rules limited blanket authority to $17.5 million projects and 45-day advance notice.[66]

KEY POLICY ISSUES

The 111th Congress is overseeing the implementation of the Pipeline Safety Improvement Act of 2006 (P.L. 109-468) and pipeline security provisions in the Implementing Recommendations of the 9/11 Commission Act of 2007 (P.L. 110-53). It is also considering several new legislative proposals, further discussed below. In its ongoing oversight of federal pipeline safety and security activities, Congress may examine a number of key issues which have drawn particular attention in policy debate.

Pipeline Damage Prevention

According to PHMSA statistics, excavation damage was the single greatest cause of accidents among all pipeline systems between 1989 and 2008.[67] Some policy makers have proposed the establishment of federal civil penalties for violations of state "one-call" notification programs to prevent excavation damage to underground pipelines. Other stakeholders have argued that such enforcement is best performed by state regulators responsible for administering one-call programs rather than by the federal government, unless the federal government determines that a state's enforcement efforts are ineffective.[68] Consistent with this approach, P.L. 109-468 prohibits federal enforcement in states already imposing such penalties (§ 2). The act also authorizes grants to states (and certain municipalities) for improving damage prevention programs if the states have been certified (under 49 U.S.C. § 60105-60106) or can demonstrate that they are establishing an "effective" program, as subsequently defined (§ 2).

On October 29, 2009, PHMSA issued an Advanced Notice of Proposed Rulemaking (ANOPR) seeking comments on proposed regulations which would establish criteria for determining adequate state enforcement of pipeline damage prevention laws.[69] According to one legal analysis, this ANOPR "is the first step in a rulemaking process that could lead to the implementation of new civil penalty enforcement authority against excavators who may cause damage to pipelines" in the absence of enforcement action by the state where the events occurred. Comments on the ANOPR were due by December 14, 2009. The agency has not yet issued a final rule. The timing and effectiveness of PHMSA's new regulations may be an oversight consideration for Congress.

Low-Stress Pipeline Regulations

Pipelines operated at less than 20% of the specified minimum strength of the material from which they are constructed are classified as "low-stress" pipelines under 49 C.F.R. § 195.2. The regulation of low-stress pipeline regulations has come under greater Congressional scrutiny since March 2006, after a spill from a BP oil pipeline led to the partial shutdown of the Prudhoe Bay area oil field on the North Slope of Alaska. In its March 15, 2006, Corrective Action Order, PHMSA found that BP's pipelines met the definition of a "hazardous pipeline facility" under 49 U.S.C. § 60112(a), which grants general authority under the statute, but that specific federal pipeline safety regulations under 49 C.F.R. § 195 did not apply at that time because BP's pipelines were classified as "low-stress" and fell under the exception in 49 C.F.R. § 195.1(b)(3).[70]

P.L. 109-468 required PHMSA to promulgate final regulations for low-stress hazardous liquids pipelines by December 31, 2007 (§ 4). The agency issued the final regulations effective July 3, 2008.[71] In overseeing the implementation of PHMSA's final criteria for low-stress pipeline regulation, Congress may consider the balance between the potential safety benefits and the potential costs of stricter safety programs in light of BP's pipeline problems and potential problems among similar pipeline systems elsewhere in the United States.

PHMSA Safety Enforcement

The adequacy of the PHMSA's enforcement strategy has been an ongoing concern of Congress, particularly after the fatal pipeline accidents in Washington and New Mexico. Provisions in the Pipeline Safety Improvement Act of 2002 (P.L. 107-355) put added scrutiny on the effectiveness of the agency's enforcement strategy and assessment of civil penalties (§ 8). In March 2006 testimony before Congress, the Government Accountability Office (GAO) reported that PHMSA had adopted measures that appeared to be responsive to earlier concerns about enforcement, although the GAO had not reviewed the strategy or its implementation in depth.[72]

In April 2006, PHMSA testified before Congress that the agency had institutionalized a "tough-but-fair" approach to enforcement, "imposing and collecting larger penalties, while guiding pipeline operators to enhance higher performance."[73] According to the agency, $4.6 million in proposed civil penalties in 2005 was three times greater than penalties proposed in 2003, the first year higher penalties could be imposed under P.L. 107-355 (§ 8(a)).[74] P.L. 109-468 required the agency to issue monthly summaries of PHMSA enforcement actions including violation and penalty information for each action, and provide a mechanism for pipeline operators to make response information available to the public (§ 6). To meet these requirements, PHMSA has established an Internet portal with pipeline safety enforcement information.[75] Proposed penalties totaled $6.5 million in 2009 including a reported $2.3 million fine—the largest single fine in the agency's history—levied against El Paso Corporation stemming from a 2006 fatal pipeline accident near Cheyenne, Wyoming.[76]

TSA and PHMSA Cooperation

Congress has long raised questions about the appropriate responsibilities and division of pipeline security authority between PHMSA and TSA.[77] Both PHMSA and TSA have played important roles in the federal pipeline security program, with TSA the designated lead agency since 2002. In 2004, the DOT and DHS entered into a memorandum of understanding (MOU) concerning their respective security roles in all modes of transportation. The MOU notes that DHS has the primary responsibility for transportation security with support from the DOT, and establishes a general framework for cooperation and coordination. The MOU states that "specific tasks and areas of responsibility that are appropriate for cooperation will be documented in annexes ... individually approved and signed by appropriate representatives of DHS and DOT."[78] On August 9, 2006, the departments signed an annex "to delineate clear lines of authority and responsibility and promote communications, efficiency, and nonduplication of effort through cooperation and collaboration between the parties in the area of transportation security."[79]

In January, 2007, the PHMSA Administrator testified before Congress that the agency had established a joint working group with TSA "to improve interagency coordination on transportation security and safety matters, and to develop and advance plans for improving transportation security," presumably including pipeline security.[80] According to TSA, the working group developed a multi-year action plan specifically delineating roles, responsibilities, resources and actions to execute 11 program elements: identification of critical infrastructure/key resources and risk assessments; strategic planning; developing regulations and guidelines; conducting inspections and enforcement; providing technical support; sharing information during emergencies; communications; stakeholder relations; research and development; legislative matters; and budgeting.[81]

P.L. 109-468 required the DOT Inspector General (IG) to assess the pipeline security actions taken by the DOT in implementing its 2004 MOU with the DHS (§ 23). The Inspector General published this assessment in May 2008. The IG report stated that

> PHMSA and TSA have taken initial steps toward formulating an action plan to implement the provisions of the pipeline security annex.... However, further actions need to be taken with a sense of urgency because the current situation is far from an "end state" for enhancing the security of the Nation's pipelines.[82]

The report recommended that PHMSA and TSA finalize and execute their security annex action plan, clarify their respective roles specifically in LNG facility security, and jointly develop a pipeline security strategy that maximizes the effectiveness of their respective capabilities and efforts.[83] According to TSA, working with PHMSA "improved drastically" after the release of the IG report; the two agencies began to maintain daily contact, share information in a timely manner, and collaborate on security guidelines and incident response planning.[84] According to the agency, TSA and PHMSA "continue to enjoy a 24/7 communication and coordination relationship in regards to all pipeline security and safety incidents."[85]

While TSA and PHMSA appear to have recently improved their cooperation under the terms of the pipeline security annex, questions remain regarding what this cooperation entails and the ongoing roles of the two agencies with respect to pipeline security. In particular, given the limited staff in TSA's pipeline security division, and the comparatively large pipeline safety staff (especially field inspectors) in PHMSA, Congress may consider whether the TSA-PHMSA pipeline security annex optimally aligns staff resources across both agencies to fulfill the nation's overall pipeline safety and security mission. H.R. 2200 would require a study reexamining the roles and responsibilities of DHS and DOT with respect to pipeline security (§ 406).

Pipeline Security Regulations

As noted earlier in this report, federal pipeline security activities to date have relied upon voluntary industry compliance with PHMSA security guidance and TSA security best practices. By initiating this voluntary approach, PHMSA sought to speed adoption of security measures by industry and avoid the publication of sensitive security information (e.g., critical asset lists) that would normally be required in public rulemaking.[86] Provisions in P.L. 109-468 require the DOT Inspector General to "address the adequacy of security standards for gas and oil pipelines" (§ 23(b)(4)). P.L. 110-53 similarly directs TSA to promulgate pipeline security regulations and carry out necessary inspection and enforcement—if the agency determines that regulations are appropriate (§ 1557(d)). Addressing this issue the 2008 IG report states that

> TSA's current security guidance is not mandatory and remains unenforceable unless a regulation is issued to require industry compliance.... PHMSA and TSA will need to conduct covert tests of pipeline systems' vulnerabilities to assess the current guidance as well as the operators' compliance.[87]

Although TSA's FY2005 budget justification stated that the agency would "issue regulations where appropriate to improve the security of the [non-aviation transportation] modes," the agency has not done so for pipelines, and is not currently working on such regulations.[88] The pipelines industry has expressed concern that new security regulations and related requirements may be "redundant" and "may not be necessary to increase pipeline security."[89] The PHMSA Administrator in 2007 testified that enhancing security "does not necessarily mean that we must impose regulatory requirements."[90] TSA officials have questioned the IG assertions regarding pipeline security regulations, particularly the IG's call for covert testing of pipeline operator security measures. They have argued that the agency is complying with the letter of P.L. 110-53 and that its pipeline operator security reviews are more than paper reviews.[91] In accordance with P.L. 110-53 (§ 1557 (b)), has been implementing a multi-year program of pipeline system inspections, including documentation of findings and follow up reviews.[92] In its oversight of potential pipeline security regulations, Congress may evaluate the effectiveness of the current voluntary pipeline security standards based on findings from the TSA's CSR reviews, pipeline inspections, and future DOT Inspector General reports.

TSA Pipelines Security Resources

Congress has been critical in the past of TSA's funding of non-aviation security activities, including pipeline activities. For example, as one Member remarked in 2005, "aviation security has received 90% of TSA's funds and virtually all of its attention. There is simply not enough being done to address ... pipeline security."[93] At its current staffing level, TSA's Pipelines Security Division has limited field presence for inspections and possible enforcement of future regulations. In conducting a pipeline corporate security review, for example, TSA typically sends one to three staff to hold a three- to four-hour interview with the operator's security representatives followed by a visit to only one or two of the operator's pipeline assets.[94]

TSA's plan to focus security inspections on the largest pipeline and distribution system operators seeks to make the best use of its limited resources. However, there is concern that the agency's CSRs as currently structured may not allow for rigorous security plan verification nor a credible threat of enforcement, so operator compliance with security guidance may be inadequate, leaving the pipeline network as a whole less secure than it might be with more universal inspection and enforcement coverage. The limited number of CSRs the agency can complete in a year is also a concern. According to a 2009 GAO report, "TSA's pipeline division stated that they would like more staff in order to conduct its corporate security reviews more frequently," and "analyzing secondary or indirect consequences of a terrorist attack and developing strategic risk objectives required much time and effort."[95]

P.L. 110-53 specifically authorizes funding of $2 million annually through FY2010 for TSA's pipeline security inspections and enforcement program (§ 1557(e)). It is an open question whether $2 million annually is sufficient to enable TSA to meet congressional expectations for federal pipeline security activities.

Additional Issues

In addition to the issues mentioned above, Congress may consider several issues related to proposed legislation or otherwise raised by pipeline stakeholders.

Distribution Integrity Management

As noted earlier in this report, integrity management programs were mandated for oil transmission pipelines in 2001 and for natural gas transmission pipelines in 2003. Congress and other stakeholders have since sought to extend these regulations to natural gas distribution pipelines, such as those operated by regional natural gas utilities. Because distribution pipelines are designed and operate differently from transmission lines, PHMSA has been developing approaches to structuring unique regulations for distribution systems. Natural gas distribution companies have sought flexible, risk-based options in any future integrity management regulations directed at distribution systems.[96] P.L. 109-468 mandated the promulgation by PHMSA of minimum standards for integrity management programs for distribution pipelines by December 31, 2007 (§ 9). PHMSA issued final regulations requiring operators of natural gas distribution pipelines to adopt integrity management programs similar to existing requirements for gas transmission pipelines on December 3, 2009.[97] As PHMSA's implementation of distribution integrity management continues, Congress may pay particular attention to whether compliance requirements balance the potential benefits of improved pipeline safety with the potential costs to distribution pipeline operators.

Mandatory Pipeline Assessment Intervals

The Pipeline Safety Improvement Act of 2002 requires that natural gas pipeline operators subject to the act perform integrity management reassessments at least every seven years after an initial baseline assessment (§ 14a). Some pipeline operators believe that this reassessment interval may be too prescriptive and may not be appropriate for all pipelines. Operators argue that assessing pipelines too frequently is costly and inefficient, diverting limited safety resources from other uses with greater pipeline safety benefits.[98] Based on assessments conducted through 2005, "and the generally safe condition of gas transmission pipelines," the GAO concluded in 2006 that the seven year reassessment interval "appears to be conservative."[99] The GAO recommended that Congress permit pipeline operators to reassess gas transmission pipelines at intervals based on risk factors, technical data, and engineering analyses. The agency believed such a revision would allow PHMSA more flexibility to establish longer or shorter reassessment intervals as warranted by pipeline conditions.[100] According to PHMSA testimony in June 2008, the Secretary of Transportation has corresponded with House Energy and Commerce committee regarding the agency's plans for exempting pipeline operators from the seven year interval requirement, but this correspondence has not been released publicly.[101] PHMSA has since concurred with the GAO's recommendation for extending reassessment intervals and is reviewing its authority to do so through the grant of special permits to individual operators.[102] S. 1333 would allow pipeline integrity reassessment intervals to be changed from seven years to intervals based on "technical data, risk factors, and engineering analysis" (§ 401).

Natural Gas Distribution Excess Flow Valves

Excess flow valves are safety devices located in pipelines which can automatically shut off the pipeline flow in the event of a pipeline rupture. In this way, the valves can minimize the release of hazardous liquids or gases during a pipeline accident, thereby reducing the likelihood or severity of a fuel fire or explosion. P.L. 109-468 required PHMSA to promulgate minimum standards for natural gas distribution systems requiring the installation of excess flow valves on new gas distribution lines in single-family homes (§ 9). The agency issued final regulations for excess flow valves as part of its final rule for natural gas distribution integrity management programs on December 3, 2009.[103] Some in Congress have proposed the extension of excess flow valve requirements beyond single-family homes.[104]

TSA Pipeline Security Information Collection

In 2009, TSA proposed collecting security-related information from pipeline operators, including reports about security incidents (via the Transportation Security Operations Center). Among the types of incidents that would be reported on a voluntary basis are "theft of pipeline company vehicles, uniforms, or employee credentials" and various incidents or activities of a "suspicious" nature.[105] Some in industry have expressed concern that TSA's proposed incident reporting standards would expose them to civil liability and include "overbroad and unnecessary data categories," especially with respect to the term "suspicious," which TSA does not define.[106]

In another 2009 filing, TSA proposes new procedures for members of the public to report to the agency a problem, deficiency, or vulnerability regarding transportation security, including pipeline transportation.[107] The proposal would fulfill requirements for such a reporting mechanism in P.L. 110-53(§ 1413, 1521, and 1536). H.R. 4016 would establish a working group for the purpose of improving the collection and use of information about hazardous material transportation incidents (§ 203) and would increase PHMSA staffing by 84 employees (§ 304) to carry out its safety program. These provisions would apply to all modes of transportation, presumably including pipelines. As Congress continues its oversight of federal pipeline security and safety activities, it may seek to ensure that agencies are collecting pipeline incident information efficiently and consistently without creating an undue burden on the providers or on agency resources designated to utilize and maintain that information.

CONCLUSION

Both government and industry have taken numerous steps to improve pipeline safety and security since 2001. Although pipeline impacts on the environment remain a concern of some public interest groups, both federal government and industry representatives suggest that federal pipeline programs have been on the right track.

As oversight of the federal role in pipeline safety and security continues, questions may be raised concerning the effectiveness of state pipeline damage prevention programs, the promulgation of low-stress pipeline regulations, federal pipeline safety enforcement, the relationship between DHS and the DOT with respect to pipeline security, and particular provisions in federal pipeline safety regulation. In addition to these specific issues, Congress may wish to assess how the various elements of U.S. pipeline safety and security activity fit together in the nation's overall strategy to protect transportation infrastructure. For example, diverting pipeline resources away from safety to enhance security might further reduce terror risk, but not overall pipeline risk, if safety programs become less effective as a result. Pipeline safety and security necessarily involve many groups: federal agencies, oil and gas pipeline associations, large and small pipeline operators, and local communities. Reviewing how these groups work together to achieve common goals could be an oversight challenge for Congress.

End Notes

[1] Parts of this report were previously published in CRS Report RL31990, *Pipeline Security: An Overview of Federal Activities and Current Policy Issues*, by Paul W. Parfomak.

[2] In this report "oil" includes petroleum and other hazardous liquids such as gasoline, jet fuel, diesel fuel, home heating oil, propane, and butane unless otherwise noted. Other hazardous liquids transported by pipeline include anhydrous ammonia, carbon dioxide, kerosene, liquefied ethylene, and some petrochemical feedstocks.

[3] Pipeline and Hazardous Materials Safety Admin., "Natural Gas Transmission, Gas Distribution, and Hazardous Liquid Pipeline Annual Mileage," Web table, January 12, 2010, http://www.phmsa.dot.gov/pipeline/library/data-stats.

[4] Richard A Rabinow, "The Liquid Pipeline Industry in the United States: Where It's Been, Where It's Going," Prepared for the Association of Oil Pipe Lines, April 2004, p. 4.

[5] Energy Information Administration, "Estimated Natural Gas Pipeline Mileage in the Lower 48 States, Close of 2008," Online table, 2009. http://www.eia.doe.gov/pub/oil_gas/natural_gas/analysis_publications/ngpipeline/mileage.html

[6] Pipeline and Hazardous Materials Safety Admin., January 12, 2010.

[7] Energy Information Administration, "About U.S. Natural Gas Pipelines," June 2007, pp. 1, 29. http://www.eia.doe.gov/pub/oil_gas/natural_gas/analysis_publications/ngpipeline/fullversion.pdf

[8] BTS, September 2008.

[9] Michelle M. Foss, "Introduction to LNG," Center for Energy Economics, University of Texas at Austin, January 2007, p. 5.

[10] Pipeline and Hazardous Materials Safety Administration, "Significant Pipeline Incidents," Web page, January 25, 2010. http://primis.phmsa.dot.gov/comm/reports/safety/SigPSI.html

[11] Ibid.

[12] National Transportation Safety Board, *Pipeline Accident Report* PAR-03-01, February 2003.

[13] See, for example: Bellingham Herald Editorial Board, "Citizen Need Panel To Monitor Pipeline Safety," *Bellingham Herald* (WA), Januray 24, 2010; Janet Zink, "Fueling the Resistance," St. Petersburg Times, December 16, 2007; W. Loy, "Slope Mayor Questions Leak Detection," *Anchorage Daily News*, March 14, 2006; J. Nesmith and R. K. M. Haurwitz, "Pipelines: The Invisible Danger," *Austin American-Statesman*, July 22, 2001.

[14] J.L. Shreeve, "Science & Technology: The Enemy Within," *The Independent*. London, UK, May 31, 2006, p. 8.

[15] Government Accountability Office (GAO), *Security Assistance: Efforts to Secure Colombia's Caño Limón-Coveñas Oil Pipeline Have Reduced Attacks, but Challenges Remain*, GAO-05-971, September 2005, p. 15; Stratfor Forecasting, Inc.," Colombia: The FARC's Low-Level Pipeline Campaign," *Stratfor Today*, June 23, 2008. http://www.stratfor.com/analysis/colombia_farcs_low_level_pipeline_campaign?ip_auth_redirect=1

[16] President's Commission on Critical Infrastructure Protection, *Critical Foundations: Protecting America's Infrastructures*, Washington, DC, October 1997.

[17] K. Houreld, "Militants Say 3 Nigeria Pipelines Bombed," *Associated Press*, May 8, 2007.

[18] Reed Johnson, "Six Pipelines Blown Up in Mexico," *Los Angeles Times,* September 11, 2007. p A-3.

[19] U.S. Dept. of Justice, "Four Individuals Charged in Plot to bomb John F. Kennedy International Airport," Press release, June 2, 2007.

[20] Elise Stolte, "EnCana Puts Record $1M on Bomber's Head," *Edmonton Journal*, July 31, 2009.

[21] "Already Hard at Work on Security, Pipelines Told of Terrorist Threat," *Inside FERC*, McGraw-Hill Companies, January 3, 2002.

[22] Alyeska Pipeline Service Co., Internet page, Anchorage, AK, February 8, 2009. http://www.alyeska-pipe.com/about.html.

[23] D. S. Cloud, "A Former Green Beret's Plot to Make Millions Through Terrorism," *Ottawa Citizen*, December 24, 1999, p. E15.

[24] Y. Rosen, "Alaska Critics Take Potshots at Line Security," *Houston Chronicle*, February 17, 2002.

[25] W. Loy, "Web Post Urges Jihadists to Attack Alaska Pipeline," *Anchorage Daily News*, January 19, 2006.

[26] U.S. Attorney's Office, Middle District of Pennsylvania, "Man Convicted of Attempting to Provide Material Support to Al-Qaeda Sentenced to 30 Years' Imprisonment," Press release, November 6, 2007; A. Lubrano and J. Shiffman, "Pa. Man Accused of Terrorist Plot," *Philadelphia Inquirer*, February 12, 2006, p. A1.

[27] Safety and security of liquified natural gas (LNG) facilities used in gas pipeline transportation is regulated under CFR Title 49, Part 193.

[28] U.S. Office of Management and Budget, *Budget of the United States Government, Fiscal Year 2011: Appendix*, February 2010, p. 989.

[29] 49 U.S.C. 601. States may recover up to 50% of their costs for these programs from the federal government.

[30] U.S. Office of Management and Budget, *Budget of the United States Government, Fiscal Year 2010: Appendix*, February 2009, p. 952.

[31] U.S. Office of Management and Budget, February 2010, p. 988.

[32] Pipeline and Hazardous Material Safety Administration (PHMSA), "PHMSA Pipeline Safety Program: Summary of Enforcement Actions," Web page, January 25, 2010. http://primis.phmsa.dot.gov/comm/reports/enforce/Actions_opid_0.html

[33] Research and Special Programs Administration (RSPA), *Pipeline Safety. Pipeline Integrity Management in High Consequence Areas (Hazardous Liquid Operators with 500 or More Miles of Pipeline), Federal Register*, December 1, 2000, p. 75378.

[34] P.L. 107-355 encourages the implementation of state "one-call" excavation notification programs (§ 2) and allows states to enforce "one-call" program requirements. The act expands criminal responsibility for pipeline damage to cases where damage was not caused "knowingly and willfully" (§ 3). The act adds provisions for ending federal-state pipeline oversight partnerships if states do not comply with federal requirements (§ 4).

[35] A 2006 Government Accountability Office (GAO) report found that PHMSA's gas integrity management program benefitted public safety, although the report recommended revisions to PHMSA's performance measures. See GAO. Natural Gas Pipeline Safety: Integrity Management Benefits Public Safety, but Consistency of Performance Measures Should Be Improved, GAO-06-946, September 8, 2006, pp. 2-3.

[36] Presidential Decision Directive 63, *Protecting the Nation's Critical Infrastructures*, May 22, 1998.

[37] Research and Special Programs Administration (RSPA), *RSPA Pipeline Security Preparedness*, December 2001.

[38] Ellen Engleman, Administrator, Research and Special Programs Administration (RSPA), statement before the Subcommittee on Energy and Air Quality, House Energy and Commerce Committee, March 19, 2002.

[39] Ellen Engleman, Administrator, Research and Special Programs Administration (RSPA), statement before the Subcommittee on Highways and Transit, House Transportation and Infrastructure Committee, February 13, 2002.

[40] James K. O'Steen, Research and Special Programs Administration (RSPA), *Implementation of RSPA Security Guidance*, presentation to the National Association of Regulatory Utility Commissioners, February 25, 2003.

[41] Office of Pipeline Safety (OPS), personal communication, June 10, 2003.

[42] HSPD-7 supersedes PDD-63 (par. 37).

[43] Executive Order 13416, "Strengthening Surface Transportation Security," December 5, 2006.

[44] These offices were formerly known as the Pipeline Security Program Office and the Intermodal Security Program Office, respectively.

[45] Department of Transportation (DOT), "Action Taken and Actions Needed to Improve Pipeline Safety," CC-2004-061, June 16, 2004, p. 21.

[46] Transportation Security Administration, Personal communication, February 2, 2010.

[47] Ibid.

[48] Mike Gillenwater, TSA, "Pipeline Security Overview," Presented to the Alabama Public Service Commission Gas Pipeline Safety Seminar, Montgomery, AL, December 11, 2007.

[49] TSA, *TSA Multi-Modal Criticality Evaluation Tool*, TSA Threat Assessment and Risk Management Program, slide presentation, April 15, 2003.

[50] TSA, Intermodal Security Program Office, *Pipeline Security Best Practices*, October 19, 2005, p. 1.

[51] Transportation Security Administration, Personal communication, February 2, 2010.

[52] Ibid.

[53] U.S. Office of Management and Budget, *Budget of the United States Government, Fiscal Year 2011: Appendix*, February 2010, p. 526.

[54] Transportation Security Administration, Pipeline Security Division, personal communication, February 2, 2010.

[55] Hawley, Kip, Asst. Secretary, Dept. of Homeland Security, Testimony before the Senate Committee on Commerce, Science, and Transportation hearing on Federal Efforts for Rail and Surface Transportation Security, January 18, 2007.

[56] Transportation Security Administration, personal communication, February 2, 2010.

[57] Government Accountability Office, *GAO Watchdog*, "Transportation Security's Efforts To Ensure Pipeline Security," Assignment No. 440768, Internet database, February 4, 2010.

[58] Pipeline and Hazardous Materials Safety Administration (PHMSA), "Pipeline Safety: Lessons Learned From a Security Breach at a Liquefied Natural Gas Facility," Docket No. PHMSA-04-19856, *Federal Register*, Vol. 71, No. 249, December 28, 2006, p. 78269; TSA, Intermodal Security Program Office, personal communication, August 30, 2006.

[59] TSA, July 6, 2007.

[60] U.S. Code of Federal Regulations. 18 C.F.R. 157.

[61] Federal Energy Regulatory Commission (FERC), personal communication, May 22, 2003.

[62] Federal Energy Regulatory Commission (FERC), News release, R-01-38, Washington, DC, September 14, 2001.

[63] Federal Energy Regulatory Commission (FERC), *Federal Energy Regulatory Commission Annual Report FY2005*, 2006, p. 19. These are the most recent specific figures reported.

[64] Federal Energy Regulatory Commission (FERC), *Federal Energy Regulatory Commission* Annual Report FY2006, 2007, p. 23.

[65] Federal Energy Regulatory Commission (FERC), News release, R-03-08, Washington, DC. February 20, 2003.

[66] Schmollinger, Christian, "FERC OKs Emergency Reconstruction," *Natural Gas Week*, May 13, 2003.

[67] Pipeline and Hazardous Materials Safety Admin., "Damage Prevention," Web page, October 14, 2009, http://primis.phmsa.dot.gov/comm/DamagePrevention.htm?nocache=403.

[68] Felt, T., President and CEO, Explorer Pipeline, Statement before the House Committee on Energy and Commerce, Subcommittee on Energy and Air Quality hearing on Reauthorization of the Pipeline Safety Act, July 27, 2006.

[69] Department of Transportation, Pipeline and Hazardous Material Safety Administration, "Pipeline Safety: Pipeline Damage Prevention Programs," 74 *Federal Register* 55798, October 29, 2009.

[70] Pipeline and Hazardous Material Safety Admin. (PHMSA), *Corrective Action Order in the Matter of BP Exploration (Alaska), Inc., Respondent*, CPF No. 5-2006-5015H, March 15, 2006.

71 Pipeline and Hazardous Materials Safety Administration (PHMSA), "Protecting Unusually Sensitive Areas From Rural Onshore Hazardous Liquid Gathering Lines and Low-Stress Lines: Final Rule," PHMSA—RSPA—2003—15864, *Federal Register*, Vol. 73, June 3, 2008. p. 31634.

72 Siggerud, K. Government Accountability Office (GAO), Testimony before the House Committee on Transportation and Infrastructure, Subcommittee on Highways, Transit and Pipelines hearing on Pipeline Safety, GAO-06-474T, March 16, 2006, p. 11.

73 Gerard, S.L., Pipeline and Hazardous Materials Admin.(PHMSA), Testimony before the House Energy and Commerce Committee, Energy and Air Quality Subcommittee hearing on Pipeline Safety, Serial No. 109-84, April 27, 2006, p. 14.

74 Ibid.

75 Pipeline and Hazardous Material Safety Administration (PHMSA), "PHMSA Pipeline Safety Program: Enforcement," Web page, May 15, 2007. http://primis.phmsa.dot.gov/comm/reports/enforce/Enforcement.html

76 Pipelines and Hazardous Materials Safety Admin. (PHMSA), "Civil Penalty Cases: Nationwide," Web page, January 25, 2010. http://primis.phmsa.dot.gov/comm/reports/enforce/CivilPenalty_opid_0.html?nocache=4013; "Colorado Pipeline Company Fined 2.3 Million After Explosion," *Clean Skies News*, December 1, 2009.

77 For example, see Hon. William J. Pascrell, Jr., statement at the House Committee on Transportation and Infrastructure, Subcommittee on Highways, Transit and Pipelines, hearing on Pipeline Safety, March 16, 2006.

78 Dept. of Homeland Security (DHS) and Dept. of Transportation (DOT), *Memorandum of Understanding Between the Department of Homeland Security and the Department of Transportation on Roles and Responsibilities*, September 28, 2004, p. 4.

79 Transportation Security Admin. and Pipelines and Hazardous Materials Safety Admin., "Transportation Security Administration and Pipelines and Hazardous Materials Safety Administration Cooperation on Pipelines and Hazardous Materials Transportation Security," August 9, 2006.

80 Barrett, T.J., Administrator, Pipeline and Hazardous Materials Safety Administration (PHMSA), Testimony before the Senate Committee on Commerce, Science, and Transportation hearing on Federal Efforts for Rail and Surface Transportation Security, January 18, 2007.

81 Transportation Security Administration, Pipeline Security Division, personal communication, July 6, 2007.

82 U.S. Dept. of Transportation, Office of Inspector General, *Actions Needed to Enhance Pipeline Security, Pipeline and Hazardous Materials Safety Administration*, Report No. AV-2008-053, May 21, 2008, p. 3.

83 Ibid. pp. 5-6.

84 Transportation Security Administration, Personal communication, February 2, 2010.

85 TSA, Pipeline Security Division, personal communication, July 6, 2007.

86 GAO, *Pipeline Security and Safety: Improved Workforce Planning and Communication Needed*, GAO-02-785, August 2002, p. 22.

87 U.S. Dept. of Transportation, Office of Inspector General, May 21, 2008, p. 6.

88 Department of Homeland Security (DHS), *Transportation Security Administration Fiscal Year 2005 Congressional Budget Justification*, Washington, DC, February 2, 2004, p. 20; TSA, Pipeline Security Division, personal communication, February 17, 2009.

89 American Gas Association (AGA), American Petroleum Institute (API), Association of Oil Pipelines (AOPL), and American Public Gas Association (APGA), joint letter to members of the Senate Commerce Committee providing views on S. 1052, August 22, 2005.

90 Barrett, T.J. January 18, 2007.

91 Sammon, John, Transportation Security Administration, Testimony before the House Transportation and Infrastructure Committee, Railroad, Pipelines, and Hazardous Materials Subcommittee hearing on Implementation of the Pipeline Inspection, Protection, Enforcement, and Safety Act of 2006, June 24, 2008.

92 TSA, Pipeline Security Division, personal communication, February 17, 2009.

93 Sen. Daniel K. Inouye, opening statement before the Senate Committee on Commerce, Science and Transportation, hearing on the President's FY2006 Budget Request for the Transportation Security Administration (TSA), February 15, 2005.

94 Department of Homeland Security, "Intent to Request Approval from OMB of One New Public Collection of Information: Pipeline Corporate Security Review," 74 *Federal Register* 42086, August 20, 2009.

95 U.S. Government Accountability Office, Transportation Security: Comprehensive Risk Assessments and Stronger Internal Controls Needed to Help Infrom TSA Resource Allocation, GAO-09-492, March 2009, p. 30, http://www.gao.gov/new.items/d09492.pdf.

96 E. F. Bender, Baltimore Gas and Electric Company, testimony before the House Committee on Transportation and Infrastructure, Subcommittee on Highways, Transit and Pipelines, hearing on Pipeline Safety, March 16, 2006, p. 10.

97 U.S. Department of Transportation, "DOT Issues Much-Anticipated Rules to Enhance Pipeline Safety," Office of Public Affairs, Press release, December 3, 2009.

98 J. L. Mohn, Panhandle Energy, testimony before the House Committee on Transportation and Infrastructure, Subcommittee on Highways, Transit and Pipelines, hearing on pipeline Safety, March 16, 2006, p. 9.

[99] Government Accountability Office (GAO), *Natural Gas Pipeline Safety: Risk-Based Standards Should Allow Operators to Better Tailor Reassessments to Pipeline Threats*, GAO-06-945, September 8, 2006. p. 3.
[100] Ibid. p. 6.
[101] Gerard, Stacy, Asst. Administrator, Pipeline and Hazardous Materials Safety Administration (PHMSA), Testimony before the House Transportation and Infrastructure Committee, Railroad, Pipelines, and Hazardous Materials Subcommittee hearing on Implementation of the Pipeline Inspection, Protection, Enforcement, and Safety Act of 2006, June 24, 2008.
[102] E. Komiskey, Office of Pipeline Safety, "High Consequence Areas and Pipeline Assessment Intervals," *Pipeline Safety – What More Needs To Be Done?*, Pipeline Safety Trust Conference, New Orleans, November 20, 2008.
[103] U.S. Department of Transportation, "DOT Issues Much-Anticipated Rules to Enhance Pipeline Safety," Office of Public Affairs, Press release, December 3, 2009.
[104] Senator Frank R. Lautenberg, Statement before the Senate Committee on Commerce, Science, and Transportation, *Nominations Hearing*, 111th Cong., 1st sess., September 23, 2009.
[105] Department of Homeland Security, "Intent to Request Approval from OMB for One New Public Collection of Information: Pipeline Operator Security Information," 74 *Federal Register* 37723, July 29, 2009.
[106] Interstate Natural Gas Association of America, "Re: Intent to Request Approval from OMB of One New Public Collection of Information: Pipeline Operator Security Information," Letter to the Transportation Security Administration, September 28, 2009, http://www.ingaa.org/cms/30/9093.aspx.
[107] Department of Homeland Security, "Reporting of Security Issues," 74 *Federal Register* 43088-43091, August 26, 2009.

In: Pipeline Safety and Security
Editor: Jonathan T. Miller and Lisa M. Gurton

ISBN: 978-1-61209-147-1

Chapter 2

Pipeline Security: TSA Has Taken Actions to Help Strengthen Security, but Could Improve Priority-Setting and Assessment Processes

United States Government Accountability Office

Why GAO Did This Study

The United States depends on a vast network of pipelines to transport energy. GAO was asked to review the Transportation Security Administration's (TSA) efforts to help ensure pipeline security. This report addresses the extent to which TSA's Pipeline Security Division (PSD) has (1) assessed risk and prioritized efforts to help strengthen pipeline security, (2) implemented agency guidance and requirements of the Implementing Recommendations of the 9/11 Commission Act of 2007 (9/11 Commission Act) regarding pipeline security, and (3) measured its performance in strengthening pipeline security. GAO reviewed PSD's risk assessment process and performance measures and observed 14 PSD reviews and inspections scheduled during the period of GAO's review. Although these observations are not generalizable, they provided GAO an understanding of how PSD conducts reviews and inspections.

What GAO Recommends

GAO recommends that TSA, among other things, establish time frames for improving risk model data, document its method for scheduling reviews, develop a plan for transmitting recommendations to operators, follow up on its recommendations, include performance measures linked to objectives in its pipeline strategy, and develop more outcome measures. DHS concurred with the recommendations and discussed planned actions, but not all will fully address the recommendations, as discussed in the report.

What GAO Found

PSD identified the 100 most critical pipeline systems and developed a pipeline risk assessment model based on threat, vulnerability, and consequence, but could improve the model's consequence component and better prioritize its efforts. The consequence component takes into account the economic impact of a possible pipeline attack, but not other possible impacts such as public health and safety, as called for in the Department of Homeland Security's (DHS) risk management guidance. PSD plans to improve its model by adding more vulnerability and consequence data, but has no time frames for doing so. Establishing a plan with time frames, as called for by standard management practices, could help PSD enhance the data in, and use of, its risk assessment model. Also, PSD procedures call for scheduling Corporate Security Reviews (CSR)—assessments of pipeline operators' security planning—based primarily on a pipeline system's risk, but GAO's analysis of CSR data suggests a system's risk was not the primary consideration. Documenting a methodology for scheduling CSRs that includes how to balance risk with other factors could help PSD ensure it prioritizes its oversight of systems at the highest risk.

PSD has taken actions to implement agency guidance that outlines voluntary actions for pipeline operators and 9/11 Commission Act requirements for pipeline security, but lacks a system for following up on its security recommendations to pipeline operators. PSD established CSR and Critical Facility Inspection (CFI) Programs in 2003 and 2008, respectively, and has completed CSRs of the 100 most at-risk systems, started conducting second CSRs, and completed 224 of 373 one-time CFIs. Both programs result in recommendations, but PSD does not generally send CSR recommendations to operators in writing or follow up to ensure that CSR and CFI recommendations were implemented. Standard project management practices call for plans that define approaches and start dates and Standards for Internal Control in the Federal Government calls for monitoring to ensure review findings are resolved. Developing a plan for how and when PSD will begin transmitting CSR recommendations to operators, and following up on CSR and CFI recommendations could better inform PSD of the state of pipeline security and whether operators have addressed vulnerabilities.

PSD has taken steps to gauge its progress in strengthening pipeline security, but its ability to measure improvements is limited. In its pipeline security strategy, PSD does not include performance measures or link them to objectives, which GAO previously identified as desirable in security strategies. In addition, PSD developed performance measures, including one outcome measure to gauge its efforts to help operators reduce vulnerabilities identified in CSRs. However, the outcome measure does not link to all three of PSD's objectives and provides limited information on improvements in areas such as physical security. According to DHS risk management guidance, outcome measures should link to objectives. Including measures linked to objectives in its strategy and developing more outcome measures directly linked to all of its objectives could help PSD improve accountability and assess improvements.

ABBREVIATIONS

AGA	American Gas Association
AOPL	Association of Oil Pipe Lines
APGA	American Public Gas Association
API	American Petroleum Institute
CFI	Critical Facility Inspection
CSR	Corporate Security Review
DOE	Department of Energy
DHS	Department of Homeland Security
DOT	Department of Transportation
FBI	Federal Bureau of Investigation
FMFIA	Federal Managers' Financial Integrity Act of 1982
HSPD-7	Homeland Security Presidential Directive-7
INGAA	Interstate Natural Gas Association of America
MOU	memorandum of understanding
NIPP	National Infrastructure Protection Plan
PHMSA	Pipeline and Hazardous Materials Safety Administration
PSD	Pipeline Security Division
TSA	Transportation Security Administration
TSNM	Office of Transportation Sector Network Management

August 4, 2010

The Honorable John Rockefeller
Chairman
The Honorable Kay Bailey Hutchison
Ranking Member
Committee on Commerce, Science, and Transportation
United States Senate

The Honorable Frank R. Lautenberg
Chairman
The Honorable John Thune
Ranking Member
Subcommittee on Surface Transportation and Merchant Marine
Infrastructure, Safety, and Security
Committee on Commerce, Science, and Transportation
United States Senate

U.S. citizens and businesses depend on the continued operation of vast networks of pipelines that traverse hundreds of thousands of miles to transport energy for operating air and surface vehicles, running industrial equipment, heating homes, and generating electricity. The United States has the largest network of energy pipelines of any nation in the world. These pipelines transport nearly all the natural gas and about two-thirds of the hazardous liquids, including crude and refined petroleum products, consumed in the United States, making them a potential target to those wanting to disrupt commerce and other activities. Although attacks on U.S. pipelines have been rare—carried out, for example, by individuals with unclear motives—attacks on pipelines outside the United States by groups such as militant rebels highlight potential vulnerabilities of pipelines. For example, in Colombia, rebels attacked a major pipeline using explosives more than 600 times from 1996 through 2005, and in Nigeria, militant rebels have repeatedly attacked pipelines and oil facilities. Within the United States, a terrorist plot to attack jet fuel pipelines and storage tanks at JFK International Airport was uncovered and foiled in 2007. The same year, a U.S. citizen was convicted of attempting to provide material support to terrorists, among other things, after he tried to conspire with Al-Qaeda to blow up sections of the Trans Alaska Pipeline System and sections of the Transcontinental Pipeline System, which carries natural gas from the Gulf Coast to New York City. Such events raise concerns that attacks could occur in the United States.

Securing the nation's pipeline system is a responsibility shared by the federal government and the private sector. Prior to the terrorist attacks of September 11, 2001, the federal government's involvement in pipelines largely focused on safety, and security efforts were minimal. In November 2001, the Aviation and Transportation Security Act established the Transportation Security Administration (TSA) within the Department of Transportation (DOT) and gave TSA the lead responsibility for security in all modes of transportation, including pipeline.[1] In November 2002, the Homeland Security Act was enacted, and upon the creation of the Department of Homeland Security, TSA was transferred from DOT to DHS, where it currently resides.[2] In August 2007, the federal government enacted the Implementing Recommendations of the 9/11 Commission Act of 2007, which required the Secretary of Homeland Security, in consultation with the Secretary of Transportation, to take specific pipeline security actions.[3] Within DHS, TSA's Pipeline Security Division (PSD) leads pipeline security activities. TSA has not issued pipeline security regulations, but works with the pipeline industry to implement suggested security measures to make pipeline systems more secure. Private companies who own and operate pipeline systems are responsible for assessing their own specific security needs and incur the costs associated with implementing security measures.

Since it is not feasible to protect all assets and systems against every possible threat, DHS has called for using a risk management approach to prioritize its investments, develop plans, and allocate resources in a risk-informed way that balances security and commerce. DHS detailed this approach in its National Infrastructure Protection Plan (NIPP), which it issued in June 2006 and updated in 2009.[4]

You requested that we review TSA's efforts to help ensure pipeline security. Specifically, this report addresses the following objectives:

- To what extent has TSA's Pipeline Security Division (PSD) identified critical pipeline systems, assessed risk, and prioritized efforts, consistent with the NIPP, to help strengthen the security of hazardous liquid and natural gas pipeline systems?
- To what extent has PSD taken actions to implement agency guidance and requirements of the Implementing Recommendations of the 9/11 Commission Act of 2007 regarding the security of hazardous liquid and natural gas pipeline systems?

- To what extent has PSD measured its performance to help strengthen the security of hazardous liquid and natural gas pipeline systems and improvements in pipeline security?

To determine the extent to which PSD used a risk management process to help strengthen the security of pipelines, we reviewed PSD's efforts to identify critical pipeline systems, assess risk, and prioritize its pipeline review efforts.[5] We reviewed relevant documents, including PSD's list of the 100 most critical pipeline systems, and interviewed PSD officials about the methods they used to identify these systems.[6] We reviewed TSA assessments of threat, vulnerability, and consequence from 2003 through May 2010—such as TSA's annual pipeline threat assessment, Corporate Security Reviews (CSR) that PSD uses as a vulnerability assessment, and consequence assessments on natural gas disruptions sponsored by the Department of Energy (DOE) and PSD—and discussed these with relevant agency officials.[7] TSA characterized these as threat, vulnerability, and consequence assessments, but we did not assess the extent to which these assessment activities met the NIPP criteria for such assessments, as this was outside the scope of our work. We analyzed PSD's risk assessment model, which integrates the various assessments to develop a risk estimate and relative risk ranking for each pipeline system, and the data PSD inputs into the model. We also compared the time elapsed between PSD's first and subsequent CSRs for each pipeline system with the system's ranking based on risk to measure the strength of their relationship. Additionally, we compared the order in which PSD conducted the first Critical Facility Inspection (CFI) for each system with each system's risk ranking, and measured the strength of that relationship.[8] To assess the reliability of April 2003 through May 2010 risk assessment model data, we (1) performed testing of required data elements, (2) compared the data with other sources of information, and (3) interviewed knowledgeable agency officials. We determined that the data were sufficiently reliable for the purposes of this report. We analyzed agency guidance on risk management, including the NIPP and the Transportation Systems Sector-Specific Plan, to determine criteria for effectively implementing a risk management framework and associated best practices for conducting risk assessments, and compared these with PSD's risk management strategy.[9] We also compared PSD's approach for advancing its risk management program to standard practices in program management planning.[10]

To determine the extent to which PSD has taken actions to implement agency guidance and Implementing Recommendations of the 9/11 Commission Act of 2007 (9/11 Commission Act) requirements regarding pipeline security, we reviewed the Pipeline Security Information Circular (2002 circular)[11] and the 9/11 Commission Act and actions described in agency documents.[12] To learn more about PSD's actions, we interviewed officials from PSD and DOT as well as representatives of the major associations with ties to the pipeline industry (American Petroleum Institute, Association of Oil Pipe Lines, American Gas Association, Interstate Natural Gas Association of America, and American Public Gas Association); attended the 2008 International Pipeline Security Forum organized by PSD and Natural Resources Canada; and met with security personnel from 10 pipeline operators with headquarters or significant operations in Houston.[13] We chose Houston because it has the highest concentration of operators with systems on PSD's list of the 100 most critical pipeline systems, and those with whom we met operate about one-third of those systems. While the results of these interviews cannot be generalized to all pipeline operators and industry associations, they provided perspectives on how operators view PSD's security efforts.

Further, we accompanied PSD officials on 4 reviews of pipeline systems operated by 4 different operators and 10 inspections of critical facilities operated by 3 different operators. We observed these reviews and inspections because PSD had scheduled them while we were conducting our work. These involved hazardous liquid and natural gas pipelines as well as different size operators with pipeline systems that varied in the amount of energy they carry, their relative risk ranking, and their location (we observed CSRs in four states and CFIs in three states). While the results of these observations cannot be generalized to all CSRs and CFIs or all pipeline systems and critical facilities, they provided us with an understanding of how PSD conducts these reviews and inspections, and some perspective on the security posture at different critical facilities. We also interviewed representatives of Secure Solutions International—a security and risk management consulting firm that assisted PSD in developing and carrying out CFIs—about critical facilities and the inspection process. In addition, we independently observed the exterior of 10 other critical facilities. We selected these facilities, which were located in four states and operated by 6 different operators, because of their proximity to our offices. Although the results of these observations cannot be generalized to all critical facilities, they provided us insight on security measures at additional critical facilities. We compared PSD's processes for transmitting and following up on CSR and CFI recommendations with criteria in GAO Standards for Internal Control in the Federal Government regarding recording and communicating deficiencies found during evaluations.[14] We also compared PSD's approach for advancing its process for communicating CSR recommendations to standard practices in project management.[15]

To determine the extent to which PSD measured its performance in strengthening the security of pipelines and improvements in pipeline security, we reviewed PSD's performance measures and interviewed Office of Transportation Sector Network Management and PSD officials regarding those measures, and discussed PSD's related data collection methodologies with PSD officials.[16] We analyzed TSA's national security strategy for pipeline systems—the Pipeline Modal Annex—to determine the extent to which it conformed to provisions related to goal setting and performance measurement found in Executive Order 13416: Strengthening Surface Transportation Security,[17] the NIPP, the Transportation Systems Sector-Specific Plan,[18] and guidance on desirable characteristics for a national strategy that we developed in a previous report.[19] We also reviewed the NIPP and the 2007 Transportation Systems Sector-Specific Plan to determine the risk management framework's recommended approach to performance measurement and compared TSA's actions with that guidance. In addition, we analyzed data PSD used as an outcome measure to determine the extent of improvements in pipeline security and evaluated both the reliability of the data and its sufficiency as a measure of pipeline security outcomes. As part of this analysis, we compared two successive data collection instruments—the original instrument PSD developed in 2003 and used in conducting early CSRs with the one TSA developed in 2004, which PSD subsequently used. Later in this report we discuss concerns about the reliability of some of these data. Appendix I contains a more detailed discussion of our objectives, scope and methodology.

We conducted this performance audit from November 2008 to August 2010 in accordance with generally accepted government auditing standards. Those standards require that we plan and perform the audit to obtain sufficient, appropriate evidence to provide a reasonable basis for our findings and conclusions based on our audit objectives. We believe that the evidence obtained provides a reasonable basis for our findings and conclusions based on our audit objectives.

BACKGROUND

Overview of U.S. Pipeline Systems

More than 2.4 million miles of hazardous liquid and natural gas pipeline—primarily buried underground in the continental United States—run under remote and open terrain as well as densely populated areas. These pipelines are comprised of three main types:

- **Hazardous liquid:** About 170,000 miles of hazardous liquid pipeline transport crude oil, diesel fuel, gasoline, jet fuel, anhydrous ammonia, and carbon dioxide.
- **Natural gas transmission and storage:** Over 320,000 miles of pipeline—mostly interstate—transport natural gas from sources to communities.
- **Natural gas distribution:** About 1.9 million miles of pipeline—mostly intrastate—transport natural gas from transmission pipelines to residential, commercial, and industrial customers.

The network of hazardous liquid and natural gas transmission pipelines in the United States can be seen in figure 1.

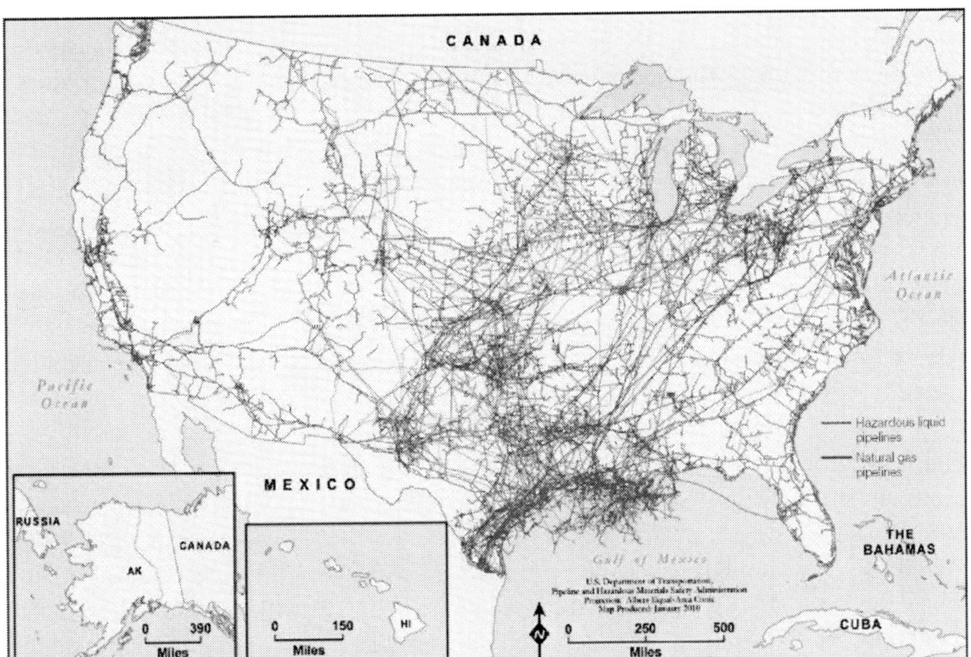

Source: Department of Transportation, Pipeline and Hazardous Materials Safety Administration.

Figure 1. Map of Hazardous Liquid and Natural Gas Transmission Pipelines in the United States, September 28, 2009.

More than 3,000 pipeline companies operate the nation's pipeline systems. Pipeline systems are comprised of the pipelines themselves, which can traverse multiple states and U.S. borders with Canada and Mexico, as well as a variety of facilities, such as storage tanks, compressor stations, and control centers. Some of these facilities are considered critical and merit particular attention to security if, for example, they are important to the nation's energy infrastructure; serve installations critical to national defense; or, if attacked, have the potential for mass casualties or significant impact on public drinking water affecting a major population center. A significant disruption of pipeline service has the potential to inflict economic havoc on a region or the nation at large.

Notwithstanding the potential damage or harm that could result from an attack, the inherent design and operation of U.S. pipeline systems might reduce some of the potential impacts regarding loss of service. For one thing, the pipeline sector is generally considered to be resilient. Historically, pipeline operators have been able to quickly respond to the adverse consequences of an incident—whether it is damage from a major hurricane or a backhoe—and quickly restore pipeline service. In addition, pipeline infrastructure is versatile and includes such redundancies as parallel pipelines or looping capabilities that enable operators to mitigate potential disruptions by rerouting energy through the network.

Key Pipeline Security Stakeholder Roles and Responsibilities

Protecting the nation's pipeline systems is a responsibility shared primarily by the federal government and private industry. Since the terrorist attacks of September 11, 2001, the role of federal agencies in securing the nation's transportation systems has continued to evolve. In response to those attacks, the federal government enacted the Aviation and Transportation Security Act of 2001, which created and conferred upon TSA broad responsibility for securing all modes of transportation, including pipeline.[20] In November 2002, the federal government enacted the Homeland Security Act, which established DHS, transferred TSA from DOT to DHS, and assigned DHS responsibility for protecting the nation from terrorism, including securing the nation's transportations systems.[21]

Within TSA, the Office of Transportation Sector Network Management (TSNM) manages all surface transportation security issues with divisions dedicated to each surface mode of transportation, including pipeline. Within TSNM, the Pipeline Security Division (PSD)—the smallest of TSNM's surface transportation divisions—has lead responsibility for the security of the nation's pipeline systems.[22] For fiscal year 2010, PSD has an authorized staffing level of 13 and a budget of about $4 million. TSA's Office of Intelligence is responsible for collecting and analyzing threat information related to the transportation network; it shares with PSD any information related to pipeline threats or suspicious incidents.

While TSA, within DHS, was given primary responsibility for pipeline security, DOT's Pipeline and Hazardous Materials Safety Administration (PHMSA) retained responsibility and authority for regulating the transportation of hazardous materials via pipeline and pipeline safety. In 2004, DHS and DOT entered into a memorandum of understanding (MOU) delineating the agencies' roles and responsibilities with respect to transportation security and recognizing DHS as having primary responsibility for security in all modes of transportation, including pipeline. In 2006, TSA and PHMSA completed an annex to the MOU further clarifying both agencies' roles. The annex identifies TSA as the lead federal entity for transportation security, including hazardous materials and pipeline security, and PHMSA as responsible for administering a national program of safety in natural gas and hazardous liquid pipeline transportation, including identifying pipeline safety concerns and developing uniform safety standards. However, pipeline security and safety are intertwined, and PSD and PHMSA coordinate on matters relating to pipeline security and protection. TSA and DOE also work together on matters where pipeline safety and security overlap and PSD and DOE worked closely on pipeline security issues, programs, and activities, such as efforts to enhance reliability and resiliency.

Although PSD has primary federal responsibility for pipeline security, implementation of asset-specific protective security measures remains the responsibility of pipeline operators in the private sector. Particularly since the September 11, 2001, terrorist attacks, operators' attention to security has increased and they have sought to incorporate security practices and programs into their overall business operations. Pipeline operators' interests and concerns are represented by five major trade associations with ties to the pipeline industry—the Interstate Natural Gas Association of America (INGAA), American Gas Association (AGA), American Public Gas Association (APGA), American Petroleum Institute (API), and Association of Oil Pipe Lines (AOPL). These associations have worked closely with the federal government on a variety of pipeline security-related issues.

In March 2002, API developed Security Guidelines for the Petroleum Industry and in September 2002, INGAA and AGA developed Security Guidelines for the Natural Gas Industry, which were adopted by APGA.[23]

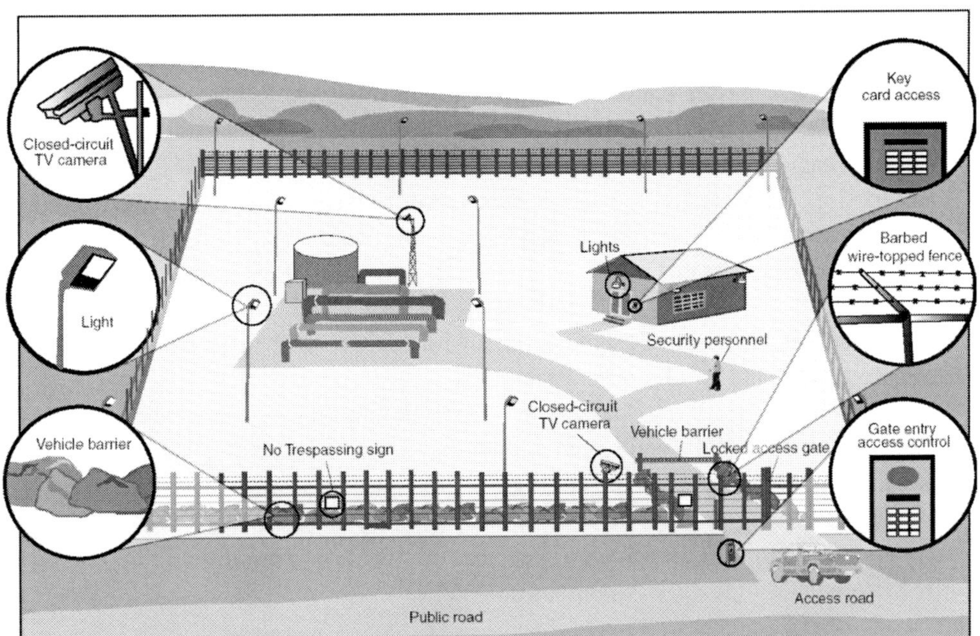

Sources: GAO analysis of Security Guidelines for the Petroleum Industry and Security Practices Guidelines for the Natural Gas Industry and Art Explosion (clip art).

Figure 2. Physical Security Measures a Pipeline Operator Might Employ at a Critical Facility.

Both sets of guidelines emphasize security planning and strategies that, to varying degrees, include identifying, analyzing, and reducing vulnerabilities. Both reference some of the physical security measures that operators can take to protect their critical facilities, but provide caveats explaining the general nature of the described security practices and the importance of each operator determining the security measures that are appropriate for each facility. Figure 2 illustrates some of the physical security measures that operators may choose to employ at a critical facility.

Laws and Agency Guidance Concerning Pipeline Security

In September 2002, prior to the establishment of DHS, DOT issued voluntary guidance for pipeline operators in the form of the Pipeline Security Information Circular (the 2002 circular), which TSA later adopted. The 2002 circular, developed in collaboration with pipeline industry associations, recommended pipeline operators identify their critical facilities, develop security plans consistent with prior industry association guidance, and begin implementing appropriate security measures at critical facilities. It also outlined steps the federal government planned to take, including conducting onsite reviews of pipeline operators' security plans to determine whether the plans are consistent with security guidance published by their industry. In collaboration with industry associations, PSD developed new, draft pipeline security guidance to replace the 2002 circular. As of May 2010, PSD had not yet issued the new guidance, but it anticipates doing so sometime during 2010.

Pipeline Security Contingency Planning Guidance, also developed by DOT in 2002 and considered part of the 2002 circular, provides criteria for pipeline operators to use to identify critical facilities and establishes guidelines for protective measures for critical facilities under each threat condition corresponding to the Homeland Security Advisory System. For example, during periods of elevated threat conditions (yellow), operators should ensure, among many other things, that employees are educated on security standards and procedures; fencing, locks, camera surveillance, intruder alarms, and lighting are in place and functioning; gates and barriers are closed and locked except those needed for immediate entry and exit at critical facilities; and visitation is limited and it is confirmed that every visitor is expected and has a need to be at a critical facility. However, similar to industry guidelines, the Pipeline Security Contingency Planning Guidance also states that pipeline operators are expected to use good judgment in incorporating measures into their security plans as not all security measures are appropriate for all types of facilities. [24]

In August 2007, Congress passed the 9/11 Commission Act, which identifies the following pipeline security requirements that the Secretary of Homeland Security must implement. Some of these requirements are shared responsibilities with the Secretary of Transportation; others are to be carried out in consultation with the Secretary of Transportation.[25] Within DHS, PSD has responsibility for carrying out the following pipeline security requirements of the 9/11 Commission Act:

- Establish a program for reviewing pipeline operators' adoption of the 2002 circular, including the review of pipeline security plans and critical facility inspections.

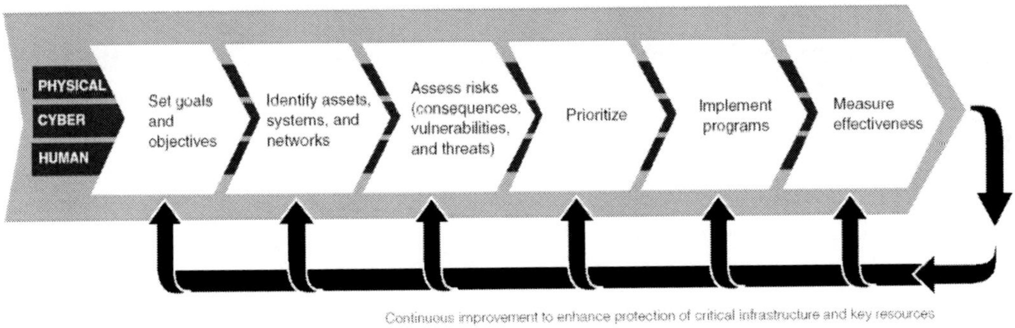

Source: GAO, DHS.

Figure 3. NIPP Risk Management Framework.

- Develop and implement a plan for reviewing the pipeline security plans of the 100 most critical pipeline operators covered by the 2002 circular.
- Develop and implement a plan for inspecting the critical facilities of the 100 most critical pipeline operators covered by the 2002 circular.
- In conducting these reviews and inspections, use risk assessment methodologies to prioritize risks and target inspections.
- Develop security recommendations for natural gas and hazardous liquid pipelines and pipeline facilities and transmit to pipeline operators.
- If the Secretary of Homeland Security determines that regulations are appropriate, promulgate regulations and carry out necessary inspection and enforcement actions.
- Develop a pipeline security and incident recovery protocols plan and submit a report to the appropriate congressional committees. The report is to include the plan and an estimate of the private and public sector costs to implement any recommendations.

A Risk-Based Approach to Guide Pipeline Security

In recent years, we, along with Congress, the executive branch, and the 9/11 Commission, have recommended that federal agencies with homeland security responsibilities utilize a risk management approach to help ensure that finite national resources are dedicated to assets or activities considered to have the highest security priority. Homeland Security Presidential Directive 7 (HSPD-7) directed the Secretary of Homeland Security to establish uniform policies, approaches, guidelines, and methodologies for integrating federal infrastructure protection and risk management activities.[26] It also called for the Secretary to produce a comprehensive, integrated national plan for critical infrastructure and key resources protection to outline national goals, objectives, milestones, and key initiatives.

In response to HSPD-7, DHS released the NIPP in June 2006, and updated it in 2009. The NIPP created a risk-based framework for the development of sector-specific agency strategic plans. In keeping with the NIPP and as required by Executive Order 13416, TSA developed the Transportation Systems Sector-Specific Plan in 2007 to document the process to be used in carrying out the national strategic priorities outlined in the NIPP. The plan contains supporting modal implementation plans for each transportation mode, including pipeline. The Pipeline Modal Annex provides information on efforts to secure pipelines, as well as TSA's overall goals and objectives related to pipeline security. The cornerstone of the NIPP is the risk management framework that entails a continual process of managing risk through six interrelated activities, as illustrated in figure 3.

- **Set goals and objectives:** Define specific outcomes, conditions, and end points for an effective risk management posture.
- **Identify assets, systems, and networks:** Develop an inventory of the assets, systems, and networks deemed to be critical, and collect information pertinent to risk management.
- **Assess risks:** Evaluate risk as a function of threat, vulnerability, and consequence. Once the three components of risk have been assessed for one or more given assets, systems, or networks, integrate them into a defensible model to produce risk estimates.
- **Prioritize:** Compare risk assessment results and establish priorities based on risk. Accord the highest priority in risk management activities to those assets, systems, or networks with the highest expected losses.
- **Implement programs:** Select appropriate actions or programs to reduce or manage the risk identified.
- **Measure effectiveness:** Use metrics and other evaluation tools to measure progress and assess the effectiveness of protection programs that have been implemented.

PSD Has Developed a Pipeline Risk Assessment Model, but Could Strengthen Data in the Model and Better Prioritize Security Reviews and Inspections

PSD Identified the Most Critical Pipeline Systems and Developed a Risk Model, but Some Model Components Could be Strengthened

PSD identified the 100 most critical pipeline systems in the United States,[27] consistent with the NIPP, and developed a pipeline risk assessment model to generate a risk score for those systems; however, some components of PSD's model are incomplete.[28] The NIPP calls for agencies to identify the most critical assets, systems, or networks within each sector, including the transportation sector, in order to collect information pertinent to risk management. PSD relied on each pipeline system's energy throughput to identify the most critical systems from more than 3,000 systems in the United States. It has since been focusing its risk management efforts on these 100 most critical systems, which, according to PSD officials, move 85 percent of all energy within the United States.

Once critical systems have been identified, the NIPP calls for agencies to assess risk as a function of threat, vulnerability, and consequence, and to integrate these individual assessments into a model to produce a risk estimate. It further requires that the consequence component of a risk assessment take into account the impact that an event or incident would have on the economy and public health and safety, among other things. PSD was the first of TSA's surface transportation modes to develop a risk assessment model that combines all three components of risk—threat, vulnerability, and consequence—to generate a risk score. PSD's pipeline risk assessment model generates a risk score for each of the 100 most critical systems and ranks them according to risk.[29] PSD holds the threat score constant for all pipeline systems and uses the results of its Corporate Security Reviews (CSR) in its vulnerability component. However, its consequence component is incomplete in that it accounts for economic impact, but not the impact on public health and safety. The following provides more information on the assessments or information for the threat, vulnerability, and consequence data that PSD uses in its risk assessment model.

- **Threat:** In the case of terrorist attacks, the NIPP calls for the threat component of the assessment to be calculated based on the likelihood of the intent and capability of a terrorist attack on a particular asset, system, or network. However, if threat likelihoods cannot be estimated, an agency can use conditional risk values based on vulnerability and consequence. TSA's Office of Intelligence develops Pipeline Threat Assessments and, according to officials from that office, the approach they use to assess threat is consistent across transportation modes. They further explained that because they have no actionable intelligence for specific pipeline systems, they can not develop likelihood estimates.[30] As such, PSD holds threat constant in its model and bases each pipeline system's risk score on vulnerability and consequence. Office of Intelligence officials explained that if they were to receive intelligence regarding a credible threat to a specific pipeline system, they would work with PSD officials to adjust the threat level for that system in PSD's risk assessment model.
- **Vulnerability:** According to the NIPP, agencies are responsible for ensuring that vulnerability assessments are performed within their sector in order to identify areas of weakness within a system under review. PSD uses the results of the CSRs it conducts on each of the most critical pipeline systems as the basis for the vulnerability component in its risk assessment model. PSD uses a CSR protocol (i.e., a questionnaire that guides the CSR interview) to collect information on an operator's security planning and management practices for a given pipeline system, and calculates a CSR score by tallying points associated with responses to each of 73 standard questions in the protocol.[31] Using the CSR score, PSD determines a pipeline system's vulnerability by calculating the difference or "gap" between a total possible score of 100 and an operator's CSR score. PSD uses this gap, known as the "vulnerability gap," as the basis for the vulnerability component in its risk assessment model. Using CSRs as vulnerability assessments is consistent with the approach taken by other surface transportation modes, such as freight rail and highway infrastructure, on which we have previously reported.[32]
- **Consequence:** According to the NIPP, consequence assessments should measure key effects on the well being of the nation. This includes the negative consequences on the economy, public health and safety, and the environment, as well as the functioning of government that can be expected if an asset, system, or network is damaged, destroyed, or disrupted by a terrorist attack. Within its risk assessment model, PSD uses the annual energy throughput of a pipeline system to help measure the possible adverse economic impact of a terrorist attack or other event on a pipeline system, but does not take into account other possible adverse impacts, such as on public health and safety. According to PSD officials, because the major consequence

of an attack on a pipeline would be the loss of energy, annual energy throughput provides a good measure of this expected loss. However, the consequences of some potential attacks might not be limited to the economy. For example, under some circumstances, an attack on a critical pipeline facility located near a waterway has the potential to significantly contaminate drinking water or, if located in a highly populated area, could result in significant casualties.

PSD officials explained that the pipeline risk assessment model is in the early stages of development and they intend to improve it over time by incorporating additional data. PSD has sponsored or conducted assessments and collected information on pipeline systems, some of which could be used to enhance individual components of its model. For example, through its Critical Facility Inspection (CFI) Program, PSD has collected information on critical facilities, such as the number of facilities per system, and officials say they plan to eventually use these data in their risk estimates. PSD officials explained that the number of critical facilities can be an indicator of a system's vulnerability—that is, the more critical facilities a system has, the more vulnerable the system. Thus, incorporating this information into the vulnerability component of PSD's risk assessment model and including it in the risk estimate could enhance the model. In addition, including information that might be available from other sources, such as the number of miles of pipeline that run through a high-consequence, or highly populated, area could also enhance the consequence component of the model. PSD officials noted that such data could be a good measure of the effects on public health and safety. However, the officials explained that with a small staff, they have not had time to make any specific enhancements to the model.

PSD officials also agreed that adding other information that could be available in the future might further improve its model. For example, PSD and DOE sponsored regional gas pipeline studies that include information that could be used to improve the consequence component of the model. These studies use computer-based modeling to evaluate the impact of a major natural gas pipeline disruption. PSD officials told us they would like to incorporate such information into the consequence component of its risk model, but adding such information for natural gas pipelines without adding comparable information for hazardous liquid pipelines would skew its risk ranking of the most critical pipeline systems. PSD officials told us in May 2010 that they had secured funds to contract for a similar assessment of the hazardous liquid pipeline market and expect the work to begin later in fiscal year 2010. They also said they plan to use the results of their CFIs to enhance the vulnerability components of the risk model; however, they will need to wait until they complete inspections of all critical facilities associated with the 100 most critical pipeline systems. The officials told us they expect to complete these inspections by the end of 2011.

Although PSD officials said they would like to add more information to their pipeline risk assessment model and have included placeholders in the model for incorporating other vulnerability and consequence factors when additional information is known, they have not established time frames or milestones (i.e., a schedule of actions needed to achieve goals) for doing this. Standard practices for program management call for establishing time frames and milestones as part of a plan to ensure that results are achieved.[33] Developing a plan that includes time frames and milestones could help PSD accomplish its goal of improving the data in its risk assessment model. By including additional information in its risk model—some that exists and some that should be available in the future—PSD could improve its risk assessment of the most critical pipeline systems and better assure it has the information it needs to guide decisions, including allocating resources to the highest risk pipeline systems. Table 1 summarizes all of TSA's assessment activities related to the three individual components of risk for the pipeline industry, and identifies which ones PSD includes in the data it inputs into its risk assessment model.[34]

Table 1. TSA Pipeline Security Assessment Activities Since 2003

Entity	Time frame	Description	Risk component addressed			Included in pipeline risk assessment model
			Threat	Vulnerability	Consequence	
TSA Office of Intelligence	Annually	**Annual Threat Assessments:** TSA's Office of Intelligence provides an overview of threats—including key actors and possible attack tactics and targets—to pipeline systems. The assessments include incidents of interest and suspicious activities targeting pipeline systems in the United States and overseas.	X			**Yes.** PSD uses this for the threat component of the risk model.
TSA Pipeline Security Division	Ongoing since 2003	**Corporate Security Reviews (CSR):** PSD conducts CSRs to assess pipeline security plans at the 100 most critical pipeline systems in the United States. The intent of these on-site reviews of pipeline companies is to develop firsthand knowledge of security planning, establish communication with key pipeline security personnel, and identify and share good security practices.		X	X[a]	**Yes.** PSD uses CSRs for the vulnerability component of the risk model.
TSA Pipeline Security Division	Ongoing since Nov. 2008	**Critical Facility Inspections (CFI):** PSD, with the help of a contractor, conducts in-depth inspections of all the critical facilities of the 100 most critical pipeline systems in the United States.		X	X	**No.** PSD collects the number of critical facilities per system, which could be used to enhance the risk model. PSD also collects consequence information for each system that could be used once PSD completes all inspections.
TSA Pipeline Security Division and Natural Resources Canada	2004-2007	**Pipeline-Cross-Border Vulnerability Assessments Program:** U.S. and Canadian teams assess pipeline operations, control systems, interdependencies, and assault planning in critical cross-border infrastructure.		X	X	**No.** PSD cannot use this in its risk model because it involves only a few of the 100 most critical pipeline systems.

Table 1. (Continued)

Entity	Time frame	Description	Risk component addressed			Included in pipeline risk assessment model
			Threat	Vulnerability	Consequence	
TSA Pipeline Security Division –initiated by DOE	2003-2008	**Regional Gas Pipeline Studies:** PSD, in coordination with DOE, sponsored a series of studies using computer-based modeling, to evaluate the impact of a major pipeline disruption.[b]			X	**No.** PSD cannot use this information until a comparable study for hazardous liquid pipelines is completed.
TSA Pipeline Security Division	Ongoing	**Cued Assessments:** When intelligence activities indicate that a pipeline operator has been under possible terrorist surveillance, PSD works with the operator to conduct vulnerability and consequence assessments to determine the existing state of security and gaps that need to be addressed. After these assessments, PSD makes recommendations on how to close the security gaps.		X	X	**No.** PSD cannot use this information because such assessments are isolated. Thus, PSD does not have such information for all of the 100 most critical pipeline systems.

Source: GAO and PSD.

[a]PSD collects consequence information during a CSR, but does not conduct a consequence assessment.

[b]INGAA and AGA funded the first in this series of studies.

PSD Could Better Prioritize Its Reviews and Inspections of Critical Pipeline Systems Based on Risk

PSD's CSR procedures call for scheduling CSRs based primarily on a pipeline system's risk ranking as determined by its risk assessment model; however, we found a weak statistical correlation between a system's risk ranking and the time elapsed between the first and subsequent CSR for a pipeline system.[35] This suggests that a system's risk ranking was not the primary consideration in scheduling these reviews. For the pipeline systems included in PSD's risk assessment model dated May 2010, PSD had conducted 54 initial CSRs of pipeline operators who operate the 100 most critical systems, and 27 second CSRs of those operating 65 of the most critical pipeline systems.[36] Figure 4 illustrates the weak correlation we found between risk ranking and time between reviews for the 27 operators with which PSD conducted a second CSR, as denoted by data points that are not clustered near or on the line of best fit.[37] If a stronger correlation existed between these variables, the data points would be clustered closer to the line of best fit.

According to CSR procedures, using a pipeline system's risk ranking when scheduling CSRs allows PSD to consider the importance of the system to the nation's transportation infrastructure and the likelihood that the system could be attacked. Similarly, according to the NIPP, the highest priority in risk management efforts should be accorded to those systems with the highest expected losses. In addition, the 9/11 Commission Act requires that risk assessment methodologies be used to prioritize risk and to target inspection actions to the highest risk pipeline assets. According to PSD officials, a pipeline system's relative risk ranking is the primary factor driving their decision of when to schedule a subsequent CSR, however, other factors, such as geographic proximity, also affect the decision. For example, in some cases PSD officials schedule a CSR for a lower risk-ranked system that might be located in the same geographic area as a higher risk-ranked system to be efficient and reduce travel time and costs.

We also found considerable variation in the time elapsed before PSD returned to conduct a second CSR. For example, our analysis of the data in PSD's risk assessment model showed:

- Within the 15 highest risk-ranked pipeline systems, the time between the first and second CSR ranged from 1 to 7 years.[38]
- For all pipeline systems, the average time elapsed between a first and second CSR was 4.8 years, regardless of the system's risk ranking.[39]
- For 5 systems that rank in the top 15 in terms of risk, approximately 6 years elapsed between a first and second CSR—more than the average time for all systems.

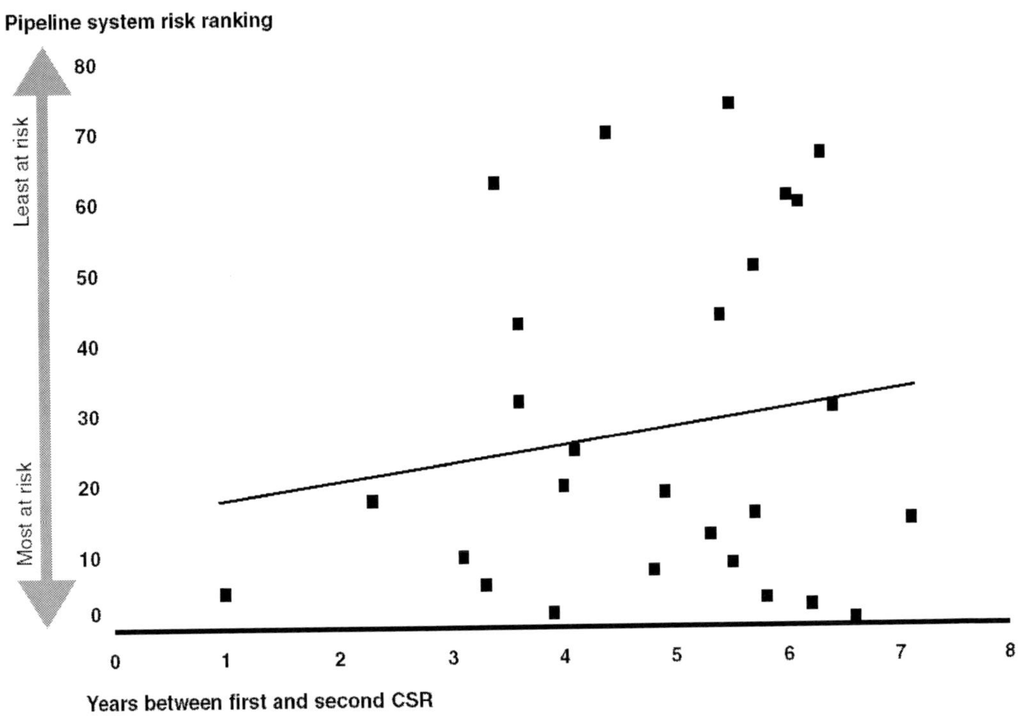

Source: GAO analysis of PSD data.

Notes: n=27.

In 27 cases, PSD conducted two CSRs for the same operator. These CSRs were conducted from April 2003 through May 2010. Because some pipeline operators operate more than one system and a CSR usually covers all the systems operated by a given operator, these 27 CSRs covered a total of 65 of the 100 most critical pipeline systems.

Figure 4. Correlation Between a Pipeline System's Risk Ranking and the Time Elapsed from the First to the Second CSR, as of May 2010.

PSD officials stated that although the time elapsed between a first and second CSR might be longer than average for some of the highest risk pipelines, this does not mean that PSD has not been focusing its attention on these operators. For example, in one of these cases, the officials explained they spent 6 weeks in 2009 inspecting dozens of critical facilities belonging to this operator through the CFI program, met with the company president to discuss the need for security improvements, and had other contacts with the operator. However, even after accounting for PSD inspecting one or more of an operator's critical facilities before conducting a second CSR, we still found a weak relationship between a pipeline system's risk ranking and the time elapsed between that system's first and subsequent CSR.[40]

The NIPP calls for systems that are considered to have the highest expected losses if damaged, disrupted, or destroyed, to receive more scrutiny. Furthermore, PSD's CSR procedures state that the CSR program should consider a pipeline system's risk level as one of the most crucial factors when scheduling CSRs and PSD officials told us they consider a system's risk to be the primary factor in these decisions. However, PSD has not clearly stated in its CSR procedures that risk should be the primary criteria in scheduling CSRs, nor has it documented a methodology addressing how it is to balance other practical considerations, such as travel efficiencies, with its consideration of risk. Doing so could help PSD ensure it prioritizes its oversight of pipeline systems that are most at risk.

Similarly, PSD has no documented procedures or methodology for using a system's risk ranking when scheduling CFIs. According to PSD officials, when they began the CFI program in November 2008, their primary consideration in scheduling CFIs was to do so in a manner that would allow them to complete a large number of inspections as soon as possible. For example, if 10 critical facilities were located close enough to each other to complete all 10 in 1 week, PSD would schedule those inspections and leave the inspections of more geographically dispersed critical facilities for a later time. The officials further explained that because inspecting outdoor space is critical to a CFI, they also consider weather when scheduling inspections (i.e., scheduling cold weather locations in warmer months). However, the NIPP calls for according the highest priority in risk management efforts to those systems with the highest expected losses. Furthermore, the 9/11 Commission Act requires that risk assessment methodologies be used to prioritize risk and to target inspections to the highest risk pipeline assets. Documenting a methodology for scheduling CFIs and including a pipeline system's risk ranking as the primary criteria while recognizing other considerations that can affect scheduling could help PSD ensure it prioritizes its oversight of pipeline systems that are most at risk.

We identified almost no statistical correlation between the order in which PSD conducted critical facility inspections and the risk ranking of the pipeline system containing those facilities.[41] For example, PSD did not inspect any of the critical facilities of three of the highest risk-ranked systems until early 2010, although it had conducted CFIs of some of the lowest risk-ranked systems in the previous year. PSD's oversight of the critical facilities belonging to the most at-risk pipeline systems could be better prioritized by scheduling inspections of facilities based on their system's risk ranking.

PSD Has Taken Actions to Implement Agency Guidance and 9/11 Commission Act Requirements, but Lacks a System for Following Up on Its Recommendations to Operators

PSD Established a Program for Reviewing Pipeline Security Plans

PSD established an on-site CSR program in April 2003 that has been evolving in response to, and consistent with, agency guidance—specifically, DOT's September 2002 Pipeline Security Information Circular (the 2002 circular)—and the 9/11 Commission Act. PSD undertook CSRs to determine the state of security within the pipeline industry and enhance the level of security planning and preparedness throughout the industry. The 2002 circular outlines voluntary actions that pipeline operators should take and describes actions the federal government plans to take to improve pipeline security.[42] It gives operators some discretion to determine which security measures are appropriate for each of their critical facilities and provides the federal government with broad guidance and, thus, some flexibility, in carrying out its reviews. According to the 2002 circular, pipeline operators should take the following actions:

- Identify critical facilities.[43]
- Develop a corporate security plan that is consistent with voluntary security guidance published by the pipeline industry.[44]
- Begin to implement appropriate security measures for the critical facilities.

In addition, the 2002 circular describes the following actions the federal government planned to take:

- Review pipeline operators' security plans on site.
- Determine whether operators' security plans are consistent with security guidance published by their industry.
- Conduct spot checks of selected critical facilities in the field to verify operators are implementing their security plans as written.
- Work with operators to correct security deficiencies.

CSRs emphasize the importance of pipeline operators' management practices in prevention, protection, and response to threats. They focus on pipeline operators' security plans and how operators manage their security programs, and include recommendations to operators for application in routine operational practices and during heightened alert levels. These reviews are also intended to provide PSD a means to establish and maintain relationships with pipeline operators' key security personnel.

CSRs include detailed interviews with the pipeline operators' security personnel—typically at operators' corporate headquarters; spot checks of selected facilities; reviews of security plans and related documents; and PSD feedback, including recommendations specific to the operator.[45] A CSR team, comprised of PSD officials, conducts the interview using a CSR protocol that PSD developed based on the 2002 circular and industry guidance.[46] The protocol includes 73 standard questions divided into 11 areas that include vulnerability assessments, credentialing, security training, cyber security,[47] and physical security.[48] According to the PSD General Manager, the CSR process gives PSD some confidence that operators are doing what their corporate security plans say. Further, he expects that operators who do well on a CSR generally have reasonably good security measures in place at their critical facilities. However, he noted that it is difficult to be certain of the physical security measures in place at critical facilities without conducting full inspections.

When the 9/11 Commission Act was enacted in August 2007, it reinforced the CSR program that PSD had underway by specifically requiring reviews of pipeline operators' security plans for the 100 most critical pipeline systems.[49] Within the first 5 years of conducting CSRs, PSD had reviewed the 100 most critical systems and had begun a second round of CSRs. As of May 2010, it had completed 103 CSRs covering more than 125 pipeline systems, including 76 first-time CSRs and 27 second-time CSRs.[50] According to PSD officials, CSRs have shown that pipeline operators are generally implementing voluntary security measures and that second CSRs have indicated that operators are generally improving their security posture.

We observed a CSR team conducting four CSRs from August through October 2009. These represented a first CSR for two of the operators and a second CSR for the remaining two operators—both of which had a first CSR in 2004. The CSR team followed the same general process for all four CSRs, asked all the questions in the CSR protocol, and conducted the CSRs in a manner consistent with CSR program goals (i.e., emphasizing the importance of security management practices, establishing working relationships with pipeline security personnel, and identifying and sharing knowledge of best practices).

The CSR team found that the security posture of these four operators varied considerably. As part of each CSR, the team identified security practices the operators were implementing well, but also made recommendations regarding areas for improvement, tailored to each operator and based on the results of each review. For the four CSRs we observed, the CSR team made a total of 32 recommendations, ranging from 3 recommendations to one operator and 17 recommendations to another. For example, officials recommended that one operator conduct vulnerability assessments for its critical facilities, another operator should issue identification cards to contractors, and a third should add certain emergency contact information to its security plan and add its new headquarters to its list of critical facilities.

PSD Established a Program for Inspecting Critical Facilities of the Most Critical Pipeline Systems

PSD established the CFI program to conduct inspections of all the critical facilities of the 100 most critical pipeline systems, as required by the 9/11 Commission Act.[51] According to PSD officials, the purpose of the CFIs is to take a one-time snapshot of each critical facility's security posture—that is, to collect information on each critical facility's security measures and equipment. PSD relied on pipeline operators to identify their own critical facilities using criteria contained in the 2002 circular. As of May 2010, operators of the 100 most critical systems had notified PSD of a total of 373 critical facilities; however, PSD officials explained that this number is fluid.

PSD manages the CFI program and has contracted with a security and risk management consulting firm that focuses primarily on energy infrastructure security to help with the program's design and implementation. CFI teams (comprised of PSD staff and contractors from the consulting firm) began conducting CFIs in November 2008 and, as of May 2010, had completed 224 CFIs. Due to the time- and resource-intensive nature of these inspections, PSD officials estimated they will finish inspecting all the critical pipeline facilities operators have identified by the end of 2011. Each CFI takes roughly 4 hours and entails the following steps:

- The CFI team conducts an in-depth interview regarding the operator's security practices using a CFI protocol that covers more than 150 items.
- The CFI team conducts an on-site physical inspection of the interior and exterior of each critical facility, including the perimeter of the property. Through physical observation and some testing, the CFI team confirms that the security measures discussed during the CFI interview are actually in place.
- The CFI team shares with the operator's security personnel its observations of good security practices, areas for improvement, and security recommendations.
- PSD sends the operator a final inspection report for each facility inspected, including recommendations, subsequent to the CFI.

From June through August 2009, we observed the CFI team conduct 10 CFIs involving critical facilities operated by three different pipeline operators. The CFI teams we observed followed the same general process for each inspection and asked all the questions in the CFI protocol. The security posture at these facilities varied considerably, and the CFI team's observations and recommendations varied accordingly. During each CFI, the team commended the operator for specific security practices that were in place at the facility, but also made recommendations for actions to improve security. The CFI team made a total of 88 recommendations for the 10 CFIs we observed, ranging from 4 recommendations at some facilities to 13 recommendations at others. For example, recommendations the CFI team made to one operator included overhauling the procedure for obtaining visitor badges and installing "no trespassing" signs and warnings indicating that the property is under video surveillance. Recommendations to another operator included securing all perimeter gates when not in active use, installing an access control system at the main gate that logs activity, upgrading main gate lighting, and establishing a formal key management program.

Figures 5, 6, and 7 show several of the security measures for which the CFI team commended the operator during a CFI we observed.

Figures 8 and 9 are photographs of two types of security lapses the CFI team identified during two other CFIs we observed. The CFI team made recommendations to the operator to address these and other security vulnerabilities.

In addition to accompanying CFI teams on inspections, we independently observed the exterior of 10 critical facilities operated by six different pipeline operators. Based on what we could observe at these facilities from outside the property perimeters, we saw variation in the physical security measures that these operators appeared to have in place—not dissimilar to what we observed when we accompanied CFI teams on their inspections.

Source: GAO.

Figure 5. Antiterrorism Crash Barrier Gate Installed inside Fenced Perimeter of a Critical Facility.

Source: GAO.

Figure 6. Boulders Installed inside Perimeter Fencing at a Critical Facility Serve as a Vehicle Barrier.

Source: GAO.

Figure 7. One of Many Closed-Circuit Television Cameras Installed at a Critical Facility.

Source: GAO.

Figure 8. CFI Team Explains That Leaving the Entry Gate of a Critical Facility Open during Business Hours Constitutes a Serious Lapse in Security.

Source: GAO.

Figure 9. Excessive Vegetation Surrounding a Critical Facility Impedes the Operator's Ability to Inspect Fencing and See Possible Intruders.

In contrast to CSRs, which look at pipeline operators' corporate security plans and security management, CFIs, when all are completed, are to yield information on security measures in place at every individual critical pipeline facility that operators have identified. According to PSD's General Manager, the CFI program fills a gap that existed in the CSR program by providing PSD the ability to develop first-hand knowledge of security measures in place at critical pipeline sites. As designed, the program provides PSD with a single point-in-time snapshot of the security posture of each critical facility. PSD officials explained that the CSR and CFI programs are complementary and that the CSRs' focus on management practices and the CFIs' focus on security measures in place at critical facilities provide PSD with needed information and are both important. They further stated that, because of its value, they are discussing ways to continue the CFI program after they complete all the inspections if resources are available. Options discussed include repeating the full set of CFIs after inspections of the critical facilities of the 100 most critical systems are completed; expanding inspections beyond these 100 systems, including toxic inhalation hazard pipeline systems; and enhancing CSRs to incorporate more thorough inspections of critical facilities.[52]

PSD Does Not Routinely Follow Up on Recommendations to Pipeline Operators

PSD does not routinely transmit its CSR recommendations in writing to pipeline operators, nor does it have a database of the CSR or CFI recommendations it makes or a process to routinely follow up on pipeline operators' implementation of those recommendations. After each CSR, PSD officials document review findings and the recommendations they make in an internal PSD report and provide oral recommendations aimed at enhancing that operator's security planning and preparedness to the pipeline operator's security personnel and sometimes management. However, PSD officials said they do not communicate these recommendations to the operator in writing as a matter of practice, but will transmit them in writing if an operator asks. Of the four CSRs we observed, one operator asked that the recommendations be put in writing, and PSD officials agreed to do so.

Standards for Internal Control in the Federal Government calls for deficiencies found during evaluations to be communicated to the individual responsible for the function and to at least one level of management above that individual. It also calls for information to be recorded and communicated to management and others within the entity who need it and in a form and within a time frame that enables them to carry out their internal control and other responsibilities. PSD officials explained they had reasons for not transmitting written recommendations to operators when they first started the CSR program, and they subsequently continued the practice of sharing recommendations orally.[53] However, by transmitting written recommendations to pipeline operators, PSD could better ensure that operators have clear guidance on actions they can take to enhance security.

PSD officials agreed that their pipeline security efforts would benefit from transmitting CSR recommendations to pipeline operators in writing and told us they intend to begin doing this after they issue new Pipeline Security Guidance and revise their CSR protocol.[54] However, they could not provide a specific time for when they would begin transmitting the recommendations to operators. Standard practices for project management call for developing a plan that includes defined approaches as well as start dates for activities.[55] Developing such a plan could help PSD accomplish its intended goal of transmitting CSR recommendations in writing to pipeline operators.

In addition, PSD officials told us they do not have a database of the recommendations they make to operators as a result of its CSRs; rather, they document CSR recommendations in individual internal reports PSD maintains on each operator. Having such a database could allow PSD to analyze the recommendations it has made through the CSR program. Moreover, the officials said they do not have a process for following up on those recommendations other than through subsequent CSRs that, on average, occur about every 5 years. According to PSD's General Manager, the greatest challenge PSD officials face is that they do not know if operators are implementing the recommendations PSD makes as a result of the CSRs. He further stated that he would like to conduct CSRs with each pipeline operator about once every 2 or 2.5 years to see if operators have implemented PSD's recommendations, but with a small staff, PSD can only visit a company about once every 4 or 5 years.[56]

Similarly, PSD officials said they do not have a database that would allow them to readily analyze the CFI recommendations they make. The CFI program, designed as a one-time inspection program of every critical pipeline facility of the 100 most critical pipeline facilities, includes recommendations that PSD sends to pipeline operators and are specific to each facility it inspects. Although the CFI contractor designed a database to capture the results of each completed CFI, the database does not include the recommendations made. Furthermore, PSD officials said they to not have a process for following up to see if operators have implemented these recommendations.

Standards for Internal Control in the Federal Government state that internal controls should generally be designed to assure that ongoing monitoring occurs, and further states that monitoring should include policies and procedures for ensuring that the findings of reviews are promptly resolved. Because PSD does not follow up on its CSR recommendations other than through a subsequent CSR 5 years later, on average, it lacks assurance that its recommendations are being implemented and whether the state of pipeline security is improving. PSD officials agreed that having a database that would allow them to analyze CSR recommendations, and following up on recommendations more frequently and systematically could increase PSD's knowledge of the security posture and vulnerabilities of individual operators as well as the pipeline industry, enhance its ability to monitor security progress, and provide additional information about its pipeline security efforts. In carrying out its CFI program, PSD has invested resources in hiring a contractor, conducting inspections, making recommendations, and developing a database. However, PSD officials agreed that without including its CFI recommendations in that database and following up on their implementation, they cannot analyze the recommendations they have made and have limited information on whether pipeline operators are addressing security vulnerabilities identified at each critical facility. PSD officials told us in May 2010 that they would like to follow up on the recommendations they make as a result of their inspections and had been discussing ways they do this, but they did not have specific plans or time frames for doing so.

Moreover, the 9/11 Commission Act states that DHS or DOT should issue pipeline security regulations if DHS determines they are appropriate. PSD officials told us in April 2009 that the results of the CSR and CFI programs, together, will inform that decision and noted that they are continually reassessing whether regulations are needed. They explained that they have been learning about the security posture of pipeline operators through these two programs and see indications that operators are making progress in improving security. Still, in a December 2009 quarterly report to the Office of Transportation Sector Network Management (TSNM) based on the first 159 CFIs, PSD reported that CFI data indicated that security improvements are needed. PSD further reported that regulations were not needed at that time. PSD officials agreed that by following up more frequently on whether operators are implementing the recommendations PSD makes as a result of its CSRs and developing a process for following up on the recommendations it makes as a result of its CFIs, they could be better informed of the state of the nation's pipeline security, including whether their recommendations have been implemented. Additionally, this would provide them information they say they plan to use to decide whether pipeline security regulations are needed.

PSD Has Developed Pipeline Security Recommendations

PSD reported that it met the 9/11 Commission Act mandate to develop and transmit security recommendations to pipeline operators through its issuance of Pipeline Security Smart Practices (Smart Practices). PSD issued its Smart Practices in August 2006 to reflect lessons learned from its first few years of conducting CSRs and to detail security practices that can enhance the security of the pipeline industry. The Smart Practices address a wide range of security practices, such as risk assessments, vulnerability assessments, and security planning; threat information; employment screening; vehicle checkpoints; physical security; intrusion detection; security awareness training; and drills, exercises, and regional cooperation. During CSRs, PSD officials remind operators of the Smart Practices and disseminate the document. In addition, PSD officials told us they inform operators of its availability through activities such as at the annual International Pipeline Security Forum and disseminate it upon request. PSD intends to periodically review and update the Smart Practices to reflect advancements in security technology and maintain the viability of the security practices described.

In addition, PSD officials stated that they will further address this mandate by issuing new Pipeline Security Guidelines to replace the 2002 circular. According to these officials, the biggest difference between the existing and new draft pipeline security guidelines is that the new voluntary guidelines will apply to all pipeline operators—including those who do not have any critical facilities. Under the new guidelines, all operators will be expected to implement some security measures at all their facilities, and implement even more at critical facilities. In contrast, the 2002 circular applies only to those operators that have critical facilities. In addition, the new guidelines will contain a section on cyber security.[57] As of May 2010, PSD officials said that the new guidelines were in draft and expected they would be issued later in 2010.

PSD officials told us they worked closely with industry groups to develop the new draft guidelines, and industry groups we spoke with commended PSD's collaborative approach during this process. An INGAA official explained that PSD used an iterative process to develop the new guidelines that included holding multiple sessions with stakeholders and forming work groups. An APGA official spoke of the open process PSD used in inviting industry comments. Similarly, AGA officials spoke highly of PSD's approach of inviting operator and association participation, which they said contributed to new guidance that applies to critical infrastructure and provides sensible baseline guidance for operators—both large and small—for securing noncritical infrastructure. API and AOPL officials also said that PSD worked closely with them and commended PSD's coordination efforts.

PSD Officials Report Developing a Pipeline Security and Incident Recovery Protocols Plan

PSD officials stated that they have drafted a pipeline security and incident recovery protocols plan, which the 9/11 Commission Act required be completed by August 2009. The 9/11 Commission Act requires that DHS develop a pipeline security and incident recovery protocols plan that includes (1) increased federal security support to the most critical pipelines under severe security threat alert levels or specific threat information and (2) a plan to develop protocols for the continued transportation of natural gas and hazardous liquids to essential markets and for essential public health or national defense uses in the event of an incident. The act required DHS to submit a report to Congress by August 2009 that included the plan and the implementation costs of any recommendations in the plan.

The plan is also to take into account actions and plans of private and public entities and consult with DOT and other stakeholders specified in the 9/11 Commission Act. The act requires DHS to develop this plan in consultation with DOT and PHMSA and in accordance with the annex to the DOT/DHS MOU, the National Strategy for Transportation Security, and HSPD-7. The 9/11 Commission Act also identifies other parties that are to be consulted as DHS develops the plan.[58] According to PSD, it consulted with the various parties called for by the act in developing its plan. Starting in December 2008, PSD, in coordination with the DOT, conducted a series of meetings and interviews with DOE, DHS's Office of Infrastructure Protection, and the Federal Bureau of Investigation (FBI).[59] PSD subsequently held two workshops (in April and May 2009) at the Johns Hopkins University Applied Physics Laboratory to discuss and review the document with additional security partners and stakeholders. PSD informed us that in developing the plan, it consulted the representatives of numerous federal agencies and agency components, as well as nonfederal organizations and industry groups.[60]

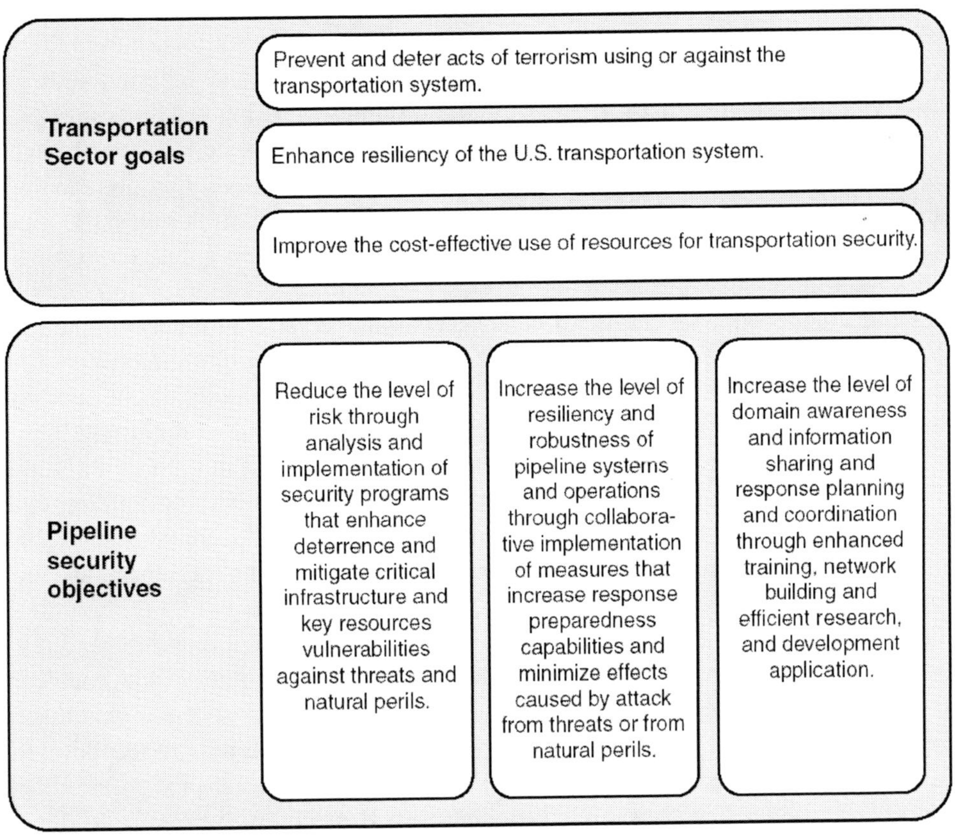

Source: GAO presentation of PSD information.

Figure 10. Transportation Sector Goals and Pipeline Security Objectives.

As of March 2010, PSD officials said they had not submitted the required report to Congress. According to the officials, the pipeline security and incident recovery protocols plan had been reviewed within DHS and was being reviewed by the Office of Management and Budget. They further said the draft plan clarifies the roles of federal agencies during and after various types of incidents, but does not contain any new responsibilities or recommendations for federal agencies or industry. As such, there are no additional costs associated with the plan and the report to Congress will not include a cost estimate.

PSD Could Strengthen Its Documented Security Strategy and More Reliably Report Security Improvements

PSD's Security Strategy Could Be Strengthened by Incorporating Performance Measures and Milestones

The 2007 Pipeline Modal Annex to the Transportation Systems Sector-Specific Plan—TSA's national security strategy for pipeline systems—identified several goals and objectives for improving transportation and pipeline security; however, the strategy lacks performance measures and milestones. In prior work, we have identified the inclusion of performance measures and milestones as a desirable characteristic for a successful national strategy and reported that a successful strategy should document what it seeks to achieve, the steps necessary to get those results, and the performance measures and milestones to gauge results.[61] We also reported that a strategy could accomplish this by stating its mission and then clearly linking its goals, objectives, programs, and performance measures to achieve results. PSD's strategy (the Pipeline Modal Annex) includes TSA's transportation sector goals that apply to all modes of transportation and identifies objectives specific to pipeline security, as shown in figure 10.[62] It also describes government and industry programs and activities that support these goals and objectives.

Although the Pipeline Modal Annex contains goals and objectives, it does not incorporate the performance measures and milestones PSD uses to evaluate the effectiveness of its security programs and activities.[63] For example, the annex describes an objective to reduce the level of risk through implementation of security programs and aligns it with the CSR program, but does not incorporate the performance measures and milestones PSD uses to evaluate the CSR program's effectiveness in achieving this objective. According to PSD officials, they considered performance measures and milestones in writing the annex, but did not include them because the annex was intended as a planning document and not an assessment tool.

Our prior work concluded that better identification of performance measures and milestones would help parties achieve results in specific time frames and enable more effective oversight and accountability.[64] Thus, using milestones and performance measures to gauge progress in meeting its stated goals and objectives could help PSD further develop and implement its national security strategy for pipeline systems and enhance its usefulness in making resource and policy decisions to better ensure accountability. Moreover, by drawing a link in the pipeline security strategy between pipeline security goals and objectives, milestones, performance measures, and programs, PSD could better evaluate its progress in helping to improve pipeline security—information that could be useful to decision makers during the risk prioritization process—and achieve results in specific time frames.

PSD Has Taken Steps to Measure Its Performance, but Could Better Measure and More Reliably Report Industry Improvements

PSD has initiated efforts to measure its performance in helping strengthen the security of pipeline systems, but could improve its performance measures to better evaluate and reliably report on the extent of security improvements in the pipeline industry. As a part of its risk management framework, the NIPP calls for agencies to measure progress in security improvements against transportation sector goals, using performance measures—(1) output data to track the progression of tasks associated with a program or activity and (2) outcome data to evaluate the extent to which a program achieves sector goals and objectives. The NIPP also states that agencies must develop performance measures that are specific and clear about what they are measuring, practical in that the needed data are available, and built on objectively measured data. NIPP Metrics Program guidance, intended to help agencies develop performance measures, called for focusing on output measures in 2008, but continuing progress toward outcome-based performance measures in 2009.

PSD Has Developed Several Performance Measures

Although the national security strategy for pipeline systems—the Pipeline Modal Annex—does not include performance measures, PSD has developed two output measures and one outcome measure to help evaluate its progress in meeting program objectives, consistent with the requirements of the NIPP. For its output measures, PSD tracks

- the number of CSRs it conducts, with a milestone, or interim goal, of 12 CSRs each year; and
- the number of CFI trips it completes, with a milestone of 15 trips each year.[65]

According to PSD officials, they track CSR and CFI program progress against these two performance milestones, and provide this information to TSNM to consider in developing the transportation sector annual report.[66]

In addition, PSD officials told us that they collect performance output data on other activities and have established the following annual milestones:

- ten stakeholder conference calls,
- an International Pipeline Security Forum,
- quarterly meetings with DOT (per PHMSA's and TSA's annex to the MOU between DHS and DOT), and
- two pipeline Intermodal Security Training Exercise Program exercises.[67]

In 2009, PSD developed an outcome measure—the vulnerability gap—that uses CSR program data to help evaluate the impact of its efforts to improve pipeline security. This outcome measure is intended to evaluate improvements in operators' security planning and preparedness based on its CSR program evaluations. More specifically, it compares the results of first and second CSRs to quantify the extent to which operators have reduced security vulnerabilities identified through CSRs.

Additional Outcome Measures Could Assist PSD in Measuring Pipeline Security Improvements

Although PSD has taken steps to gauge the progress of its programs, its ability to measure improvements in pipeline security is limited. The NIPP states that using performance measures as part of risk management can enable agencies to assess security improvements, and it instructs agencies to track progress toward a strategic goal or objective by measuring results or outcomes. The NIPP further states that the key to NIPP performance management is aligning outcome performance measures to goals and objectives.

According to the Transportation Systems Sector-Specific Plan, outcome measures should be used to assess program goals and objectives; however, output measures may be used as proxies for outcome measures in the early stages of its programs. In addition, we have reported on the limitations of output-based measures in our prior work. Specifically, we have stated that using output measures to evaluate security program performance may not systematically target areas of higher risk and may not result in the most effective use of resources because these measures are not pointed toward outcomes, or what activities are accomplishing.[68]

PSD's outcome measure—the vulnerability gap—measures aspects of two of its pipeline security objectives; however, PSD has not developed outcome measures that enable it to fully assess improvements related to pipeline security as a whole. The vulnerability gap focuses on what PSD measures through its CSR program—primarily improvements in pipeline operators' security planning and preparedness—but provides limited information on improvements in other areas, such as physical security. According to the Pipeline Modal Annex, the CSR program evaluates aspects of two of the pipeline security objectives—(1) to reduce risk and (2) to increase information sharing and response planning and coordination. By extension, the vulnerability gap measures these as well. For example, the vulnerability gap takes into account operators' risk reduction activities such as how they assess threats and vulnerabilities. It also measures increased information sharing, such as how operators manage threat information.

However, according to the Pipeline Modal Annex, the CSR program does not evaluate the third pipeline security objective—to increase the level of resiliency and robustness of pipeline systems—and, thus, the vulnerability gap does not measure this objective.[69] As a result, PSD is limited in its ability to measure or report on improvements in this latter area of pipeline security. Furthermore, according to PSD officials, collecting CSR information every 4 to 5 years limits their ability to measure the security improvements that operators are making. Nevertheless, they said the changes they have observed from operators' first to second CSRs provide them with a strong level of confidence that improvements have occurred.

PSD officials explained that they are in the early stages of performance measurement and have not yet developed additional outcome measures or established time frames for doing so. We recognize challenges PSD might face in developing outcome measures related to reducing risk. In our prior work we acknowledged that assessing the deterrent benefits of a program is inherently challenging because it is often difficult to isolate the impact of an individual program on behavior that may be affected by multiple other factors.[70] In the case of pipeline security, it may be difficult to isolate the impact of PSD's programs on operators' security actions. Nevertheless, outcome-based data could better inform decision makers of the extent to which programs and activities have been able to reduce risk and better enable them to determine funding priorities within and across agencies. Also, developing additional outcome measures that assess the impacts of its efforts to improve pipeline security and are directly aligned with transportation sector goals and pipeline security objectives could better enable PSD to evaluate security improvements in the pipeline industry.

PSD Could Improve the Reliability of Data It Uses to Measure Effectiveness

PSD designed the vulnerability gap outcome measure to help evaluate the impact of its efforts to improve pipeline security using CSR program data, but the baseline data PSD used to measure its efforts may not be reliable. When PSD officials began conducting CSRs in 2003, they developed a CSR protocol to collect information on pipeline systems' corporate security planning and preparedness. However, according to PSD officials, they began using a different protocol in August 2004 that TSA developed for all surface transportation modes to use during their respective CSRs to ensure consistency among modes. Many questions in the second protocol differed from those in the first, although the topic areas were similar.[71]

Although changes in the CSR protocol provide PSD with more information on some topics, differences between the two protocols limited PSD's ability to use CSR program data collected with the first protocol. PSD officials explained they, therefore, sought to develop comparable CSR data for all operators, regardless of which protocol PSD used during CSRs. To accomplish this, PSD officials instructed staff to reconstruct a new protocol (using the second CSR protocol) for each pipeline operator PSD reviewed from mid-April 2003 through mid-July 2004—the 15-month period during which the first protocol was used.[72] Staff were to do this using available information from the first completed protocol, any notes PSD officials took during the CSR, and security plans or other documents PSD gathered during the CSR. However, PSD officials said they did not provide written instructions to staff or verify that staff accurately reconstructed the data. Although the officials expressed confidence in their staff's work, we could not be assured that the CSR information staff reconstructed was accurate and reliable.

We analyzed the content, or substance, of the questions in both the first and second protocols and identified concerns about whether operator information could have been transferred reliably from the first to the second protocol after the fact. We found that 41 of the 73 newer CSR protocol questions were either consistent with the content of the first protocol or could have been consistently verified using the security plan operators provided during the original CSR. We therefore found it reasonable that PSD staff would have been able to accurately transfer the completed information from the first protocol to the second protocol for these 41 questions. However, we could not be reasonably assured that PSD staff accurately transferred information for the remaining 32 questions onto the second protocol because the content of these questions was inconsistent and, thus, PSD staff may not have been able to reliably reconstruct the data using the security plans operators provided during the original CSRs. For example, the second protocol contained the following questions directed to operators, but we found no similar questions on the first protocol:

- Do you have a 24/7 emergency response/operations center?
- Do you conduct different levels of background checks based on type of employment (e.g., executive, operational, police)?
- Do you periodically conduct exercises and drills?

For these questions, and 29 others like them, PSD staff may have been able to locate the information they needed in notes and documents to reconstruct the second protocol, but we had no assurance that this was possible or done in an accurate and reliable manner. We have previously reported that performance measures should reliably assess progress such that the same results would be achieved if applied repeatedly to the same situation.[73] Furthermore, errors in data accuracy could alter conclusions about the extent to which performance goals have been achieved, such as reporting performance at either a higher or lower level than is actually being attained.

We have also reported that decision makers must have assurance that the program data being used to measure performance are sufficiently accurate and reliable if the data are to inform decision-making.[74] Thus, the usefulness of agency performance information depends to a large degree on the reliability and accuracy of performance data. Because of the changes in the CSR protocol questions and concerns about the reliability of reconstructed operator responses transferred to a different form, the baseline data PSD used in comparing operators' first and second CSR scores and resulting reports may not be accurate. As such, PSD's outcome performance measure—the differences in vulnerability gaps as calculated using CSR scores—suggests a level of precision that may not be supported.[75] PSD officials said they did not see this as a significant problem because not all the baseline CSRs involved reconstructed data and, as they continue to conduct CSRs, they will eventually be able to compare the results of operators' second to third CSRs in reporting improvements. Furthermore, although PSD's CSR data may be useful for some analytical purposes, such as analyzing industry trends and assessing individual operators' security planning and preparedness, some of the early data are not useful for reporting the extent to which the vulnerability gap has closed. PSD and decision makers could be better informed and could more effectively prioritize efforts if PSD maintains a more reliable baseline for its outcome performance measure and does not use reconstructed data in reporting its baseline.

CONCLUSIONS

Securing the nation's vast network of hazardous liquid and natural gas pipeline systems is a formidable task. The importance of pipeline systems to the nation's economy underscores the need for PSD to employ a risk management approach to prioritize its security efforts. PSD has taken actions to implement a risk management approach, including identifying the 100 pipeline systems it considers most critical and being the first of the surface transportation modes to develop a risk assessment model. Nevertheless, work remains to ensure that the highest risk pipeline systems are given the necessary scrutiny. PSD's risk assessment model is in its early stages of development; however, information is available or expected that could enhance the vulnerability and consequence components of the model. By developing a plan that includes time frames and milestones for adding information to its risk assessment model,

PSD could be better assured of reaching its goal to improve the model. This could help PSD more accurately rank pipeline systems according to risk and help guide resource allocation decisions. In addition, documenting a methodology for scheduling CSRs and CFIs that includes a pipeline system's risk ranking as the primary criteria, while recognizing other considerations that can affect scheduling, could help PSD ensure it prioritizes its oversight of pipeline systems that are most at risk.

PSD has taken actions to encourage private pipeline operators to employ security measures that will protect their pipeline systems, including critical facilities. While PSD officials have said that operators of the most critical pipeline systems are generally implementing voluntary security measures, two of PSD's key efforts—its CSR and CFI programs—have identified shortcomings in operators' security programs and critical facilities that should be addressed to reduce vulnerabilities. As such, an important aspect of the CSR and CFI programs is the specific recommendations PSD makes and tailors to each operator to address the vulnerabilities PSD has identified. However, PSD is missing opportunities with respect to these recommendations. PSD officials agreed that routinely transmitting CSR recommendations in writing to operators could better ensure that operators are clear on the actions they can take to enhance the security of their pipeline system or systems, and they have said they intend to do this. Developing a plan that includes a defined approach and time frames for how and when PSD intends to begin transmitting CSR recommendations in writing to pipeline operators could help PSD accomplish its intended goal.

In addition, by establishing databases of the CSR and CFI recommendations it makes, PSD could more readily and systematically analyze its recommendations and be better informed of security vulnerabilities in the pipeline industry. Furthermore, because CSRs take place infrequently and CFIs are not repeated, following up on the implementation of CSR and CFI recommendations is particularly important. By doing so, PSD could enhance its knowledge of the state of security of the pipeline industry as well as individual systems and facilities, have an additional means for measuring the effectiveness of its programs, and obtain information that could help inform its decision on whether it would be appropriate to issue pipeline security regulations.

The 2007 Pipeline Modal Annex represents a positive step toward conveying TSA's strategy for helping the pipeline industry secure the nation's pipelines. However, incorporating PSD's performance measures and milestones and linking them to the goals and objectives in its national security strategy for pipeline systems could aid PSD and the pipeline industry in achieving results within specific time frames and could facilitate more effective oversight and accountability. PSD has developed some output-based performance measures and milestones to track the progress of its programs and activities and has developed an outcome measure to evaluate the impact of some of its efforts to improve pipeline security. However, PSD's dependence on a single outcome measure hinders its ability to evaluate the extent of improvements related to all of its pipeline security objectives. Developing additional outcome measures aligned with its objectives could facilitate PSD's efforts to better evaluate its performance.

Moreover, PSD has collected data on the security posture of pipeline operators through its CSR program and compared vulnerability gap data over time to measure the progress operators have made. PSD's CSR data may be useful to PSD for various analytical purposes. However, because of reliability issues affecting the baseline data PSD uses for calculating its vulnerability gap outcome measure, PSD would be better informed if, going forward, it establishes reliable baseline data for measuring and reporting improvements in pipeline security. Although this would limit PSD's ability to report on improvements in operators' security efforts from the first 15 months of the CSR program, it could provide greater assurance that, in the future, PSD is more accurately and reliably measuring those pipeline security improvements.

RECOMMENDATIONS FOR EXECUTIVE ACTION

To improve aspects of the Pipeline Security Division's (PSD) efforts to help ensure pipeline security, we recommend that the Assistant Secretary for the Transportation Security Administration take the following eight actions.

To ensure that PSD is managing risk effectively,

- Develop a plan with time frames and milestones for improving the data in the pipeline risk assessment model by, for example, adding more data to the consequence component.
- Document a methodology for scheduling Corporate Security Reviews (CSR) and Critical Facility Inspections (CFI) that considers a pipeline system's risk ranking as the primary scheduling criteria and balances it with other practical considerations.

To help PSD maximize its CSR and CFI efforts and keep its knowledge of the security posture of the pipeline industry current,

- Develop a plan that includes a defined approach and time frame for how and when PSD intends to begin transmitting CSR recommendations in writing to pipeline operators.
- Establish a database of CSR recommendations and develop a process for following up on the implementation of those recommendations.
- Establish a database of CFI recommendations and develop a process for following up on the implementation of those recommendations.

To better achieve the security strategy laid out in the Pipeline Modal Annex—the national security strategy for pipeline systems—to the extent feasible, revise future updates of the annex to incorporate performance measures for assessing PSD and pipeline industry progress and link those measures to pipeline security objectives.

To better evaluate PSD's performance in helping strengthen the security of hazardous liquid and natural gas pipelines and improvements in pipeline security, develop additional outcome measures that are directly linked to sector goals and modal objectives and track progress towards its stated pipeline security objectives.

To help ensure reliable reporting of security improvements in the pipeline industry, establish reliable baseline data and, until that time, refrain from using reconstructed baseline data to report progress in closing the vulnerability gap.

Agency Comments and Our Evaluation

We provided a draft of our report to DHS on July 2, 2010, for review and comment. On July 23, 2010, DHS provided written comments, which are reprinted in appendix II. In commenting on the draft report, DHS stated that it concurred with our findings and all eight recommendations and discussed efforts planned or underway to address them. However, the actions DHS reports it plans to take do not fully address the intent of four of our eight recommendations.

DHS concurred with our first recommendation that TSA develop a plan with time frames and milestones for improving the data in the pipeline risk assessment model and stated that PSD will develop a plan to coordinate security efforts that are underway that will help refine the pipeline risk ranking tool (the pipeline risk assessment model). DHS further stated that additional data from critical facility inspections, the hazardous liquid pipeline assessment, and toxic inhalation hazard study, among others, will help inform the consequence component. We support PSD's intention to develop a plan for taking such action and further encourage TSA to consider using critical facility inspection data to inform the vulnerability component of the pipeline risk model. The development of a plan for improving the data in the pipeline risk assessment model will address the intent of our recommendation, provided it includes time frames and milestones.

DHS concurred with our second recommendation that TSA document a methodology for scheduling CSRs and CFIs that considers a pipeline system's risk ranking as the primary scheduling criteria and balances it with other practical considerations. DHS stated that TSA's analysis identified as critical those pipeline systems that transport the greatest amount of energy and that PSD developed the risk ranking tool to further enhance its risk-based effort. DHS further stated that to increase the value of the risk ranking tool, PSD will develop additional data to inform the tool's rankings and base its programmatic efforts on the results. While we support PSD's intention to develop additional data to inform its ranking of pipeline systems based on risk and base programmatic efforts on those rankings, these actions, alone, will not fully address the intent of our recommendation. We believe that to better prioritize oversight of pipeline systems among the 100 that are the most critical, and to address our recommendation, TSA should document a methodology for how it will schedule pipeline CSRs and CFIs in a manner that considers risk as the primary scheduling criteria, while balancing other practical scheduling considerations, such as travel efficiencies.

DHS concurred with our third recommendation that TSA develop a plan that includes a defined approach and time frame for how and when PSD intends to begin transmitting written CSR recommendations to pipeline operators. DHS stated that PSD intends to modify its process of providing oral recommendations for security improvements to pipeline operators to include providing these recommendations to operators in writing. Developing a plan that includes a defined approach for how it will transmit its written recommendations to operators and a time frame for when it will begin to do so will address the intent of our recommendation.

DHS concurred with our fourth recommendation that TSA establish a database of pipeline CSR recommendations and develop a process for following up on the implementation of those recommendations. DHS stated that PSD will initiate the development of such a database and further stated that repeat CSRs will particularly focus on the implementation of recommendations from prior reviews. Developing a database will partially address this recommendation. However, while we support a plan that includes PSD following up on prior CSR recommendations during subsequent CSRs, this, alone, will not fully address the intent of our recommendation. Because PSD conducts a CSR for any given pipeline operator about every 5 years, on average, a process for additional and timelier follow up is needed if PSD is to be assured that its recommendations are being implemented.

DHS concurred with our fifth recommendation that PSD establish a database of CFI recommendations and develop a process for following up on the implementation of those recommendations. DHS stated that PSD has initiated the development of a CFI recommendation database and further stated that following up on those recommendations will enable TSA to assess the pipeline industry's progress in mitigating identified security deficiencies. Completing this database and developing a process for following up on the CFI recommendations will address the intent of our recommendation.

DHS concurred with our sixth recommendation that TSA revise future updates of the Pipeline Modal Annex to incorporate performance measures for assessing PSD and pipeline industry progress and link those measures to pipeline security objectives. DHS stated that in future updates to the Transportation Systems Sector-Specific Plan, PSD will include performance measures within the Pipeline Modal Annex consistent with the sector format and guidance. However, direction on what is to be included in future updates of the Pipeline Modal Annex originates with TSA, which provides transportation modes, including pipeline, with guidance and a recommended format on how to revise or rewrite modal annexes to the Transportation Systems Sector-Specific Plan. TSA's 2010 Modal Plan Revision Guidance for transportation modes does not explicitly call for incorporating performance measures for assessing modal progress and, further, linking those measures to modal objectives. Thus, without TSA direction to include performance measures that are linked to objectives in modal annex updates, the action DHS described to address our recommendation does not fully address our intent.

DHS concurred with our seventh recommendation that TSA develop additional outcome measures that are directly linked to sector goals and modal objectives and track progress towards its stated pipeline security objectives. DHS stated that PSD will develop appropriate outcome measures that reflect the impact of its security programs and the security status of the pipeline industry, and further stated that this effort will be made consistent with the performance measurement guidance of the Transportation Systems Sector-Specific Plan. We support PSD's intention to develop additional outcome measures. However, to fully address the intent of our recommendation, TSA should ensure that its performance measurement guidance calls for outcome measures to be directly linked to sector goals and modal objectives.

DHS concurred with our eighth recommendation that TSA establish reliable baseline data for reporting security improvements in the pipeline industry and, until that time, refrain from using reconstructed baseline data to report progress in closing the vulnerability gap. DHS stated that updated data from repeat CSRs will be utilized to ensure more accurate reporting of the pipeline industry's security status. Such action will address the intent of our recommendation.

DHS also provided us with technical comments, which we considered and incorporated in the report where appropriate.

As agreed with your office, unless you publicly announce the contents of the report, we plan no further distribution for 30 days from the report date. At that time, we will send copies to the Secretary of Homeland Security, the Assistant Secretary of the Transportation Security Administration, appropriate congressional committees, and other interested parties. The report also is available at no charge on the GAO Web site at http://www.gao.gov/.

If you or your staff have any further questions about this report or wish to discuss these matters further, please contact me at (202) 512-4379 or lords@gao.gov. Contact points for our Offices of Congressional Relations and Public Affairs may be found on the last page of this report. Key contributors to this report are listed in appendix III.

Stephen M. Lord
Director, Homeland Security and Justice Issues

APPENDIX I. OBJECTIVES, SCOPE, AND METHODOLOGY

Objectives

You requested that we review the Transportation Security Administration's (TSA) efforts to help ensure pipeline security. Specifically, this report addresses the following questions:

- To what extent has TSA's Pipeline Security Division (PSD) identified critical pipeline systems, assessed risk, and prioritized efforts, consistent with the National Infrastructure Protection Plan (NIPP), to help strengthen the security of hazardous liquid and natural gas pipeline systems?
- To what extent has PSD taken actions to implement agency guidance and requirements of the Implementing Recommendations of the 9/11 Commission Act of 2007 (9/11 Commission Act) regarding the security of hazardous liquid and natural gas pipeline systems?
- To what extent has PSD measured its performance to help strengthen the security of hazardous liquid and natural gas pipeline systems and improvements in pipeline security?

Scope and Methodology

To determine the extent to which PSD used a risk management process to help strengthen the security of pipelines, we reviewed PSD's efforts to (1) identify critical pipeline systems, (2) assess risk, and (3) prioritize its pipeline review and inspection efforts. To evaluate PSD's efforts to identify the most critical pipeline systems, we reviewed relevant documents, including PSD's list of the 100 most critical pipeline systems, and interviewed PSD officials about the methods they used to identify the most critical pipeline systems.

To evaluate PSD's efforts to assess risk, we reviewed TSA assessments of threat, vulnerability, and consequence that were conducted from 2003 through May 2010. Specifically, we reviewed TSA's Pipeline Threat Assessments for 2008 and 2010 and interviewed officials at TSA's Office of Intelligence. We also reviewed Corporate Security Reviews (CSR) that PSD uses as vulnerability assessments, and consequence assessments on natural gas disruptions sponsored by the Department of Energy and PSD—and discussed these assessments with relevant agency officials. TSA characterized these as threat, vulnerability, and consequence assessments, but we did not assess the extent to which these assessment activities met the NIPP criteria for threat, vulnerability, and consequence assessments, as this analysis was outside the scope of our work.

To evaluate PSD's efforts to prioritize risk, we analyzed its risk assessment model—the Pipeline Relative Risk Ranking Tool, which integrates the various assessments to produce a risk estimate and relative risk ranking for each pipeline system—and the data PSD inputs into the model. We also interviewed PSD officials about how they decide when to schedule CSRs and Critical Facility Inspections (CFI). Using correlation analysis and the data in the pipeline risk assessment model, we compared the time elapsed between PSD's first and subsequent CSR for each pipeline system with the system's ranking based on risk to measure the strength of their relationship.[76] Specifically, for those systems that had two CSRs, we assessed the strength of the correlation between the time elapsed from the first and second CSR and the system's risk ranking. We found a correlation coefficient of 0.2, which indicates a weak correlation. A correlation coefficient measures the strength and direction of linear association between two variables without controlling for the effects of other characteristics.

Because PSD officials said that the time elapsed between CSRs might be misleading because it does not account for other significant contact PSD might have had with an operator during that period, such as through a CFI, we controlled for this by running a simple regression equation.[77] Specifically, the regression equation compared the time elapsed between the first and second CSR against system risk rank and a dummy variable to denote if PSD inspected at least one critical facility belonging to an operator between the first and second CSR. This regression equation explained about 21 percent of the total variation in elapsed time between the first and second CSR. To determine the extent to which PSD prioritized the CFIs it conducted, we performed a correlation analysis to measure the strength and direction of the relationship between a system's risk ranking and the order in which PSD conducted a first CFI for that system compared with other systems. We found a correlation coefficient of 0.03, which denotes that almost no correlation exists between the two variables.

To assess the reliability of the April 2003 through May 2010 data PSD used in its risk assessment model, we (1) performed electronic testing of required data elements, (2) compared data in the model with other sources of information, and (3) interviewed agency officials knowledgeable about the data. We determined that the data were sufficiently reliable for the purposes of this report. We analyzed agency guidance on risk management, including the NIPP and the Transportation Systems Sector-Specific Plan, to determine criteria for effectively implementing a risk management framework and associated best practices for conducting risk assessments, and compared these with PSD's risk management strategy. In addition, we compared PSD's approach for advancing its risk management program to standard practices in program management planning.[78]

To determine the extent to which PSD has taken actions to implement agency guidance and 9/11 Commission Act requirements regarding pipeline security, we reviewed the Pipeline Security Information Circular (2002 circular) and the 9/11 Commission Act and actions described in agency documents. These documents included PSD's Pipeline Modal Annex, CSR Standard Operating Procedures, CSR and CFI protocols, and Pipeline Security Smart Practices.[79] To learn more about PSD's actions, we interviewed officials from PSD and DOT as well as representatives of the major associations with ties to the pipeline industry (American Petroleum Institute, Association of Oil Pipe Lines, American Gas Association, and Interstate Natural Gas Association of America, and American Public Gas Association); attended the 2008 International Pipeline Security Forum organized by PSD and Natural Resources Canada; and met with security personnel from 10 pipeline operators with headquarters or significant operations in Houston. We chose Houston because it has the highest concentration of operators with systems on PSD's list of the 100 most critical pipeline systems, and those with whom we met operate about one-third of those systems. While the results of these interviews cannot be generalized to all pipeline operators, they provided perspectives on how operators view PSD's security efforts.

To further our understanding of PSD's review and inspection processes, pipeline operators' security planning efforts, and physical security measures in place at selected critical pipeline facilities, we accompanied PSD officials on four reviews of pipeline systems operated by four different operators and 10 inspections of critical facilities operated by three different operators. We observed these reviews and inspections because PSD had scheduled them while we were conducting our work. These involved hazardous liquid and natural gas pipelines as well as different size operators with pipeline systems that varied in the amount of energy they carry, their relative risk ranking, and their location (we observed CSRs in four states and CFIs in three states). These observations further included one cross-border pipeline system and one port facility regulated under the Maritime Transportation Security Act. While the results of these observations cannot be generalized to all CSRs and CFIs or all pipeline systems and critical facilities, they provided us with an understanding of how PSD conducts these reviews and inspections, and some perspective on the security posture at different critical facilities. We also interviewed representatives of Secure Solutions International—a security and risk management consulting firm that assisted PSD in developing and carrying out CFIs—about critical facilities and the inspection process. In addition, we independently observed the exterior of 10 other critical facilities. We selected these facilities, which were located in four states and operated by six different operators, because of their proximity to GAO offices. Although the results of these observations cannot be generalized to all critical facilities, they provided us insight on security measures at additional critical facilities.

We also compared PSD's processes for transmitting and following up on CSR and CFI recommendations with criteria in the Standards for Internal Control in the Federal Government regarding the monitoring of deficiencies found during evaluations.[80] In addition, we compared PSD's approach for advancing its process for communicating CSR recommendation to standard practices in project management.[81]

To determine the extent to which PSD measured the impact of its efforts to help strengthen the security of pipelines and improvements in pipeline security, we reviewed PSD's performance measures and milestones. We analyzed TSA's national security strategy for pipeline systems—the 2007 Pipeline Modal Annex—to determine the extent to which it conformed to provisions related to goal setting and performance measurement found in Executive Order 13416: Strengthening Surface Transportation Security[82] and guidance on desirable characteristics for a national strategy that we developed in a previous report.[83] We also interviewed Office of Transportation Sector Network Management (TSNM) and PSD officials regarding PSD's performance measures and milestones and related data collection methodologies. In addition, we reviewed the 2009 NIPP and the 2007 Transportation Systems Sector-Specific Plan to determine the risk management framework's recommended approach to performance measurement and compared TSA's actions to that guidance.

To assess the reliability of the data PSD used to develop its vulnerability gap outcome measure in 2009 for reporting on the extent of improvements in pipeline security, we reviewed and analyzed related documentation and interviewed PSD officials knowledgeable about the data and PSD's data collection methods. As part of this analysis, we compared two successive data collection instruments—the original CSR protocol that PSD developed and used in conducting CSRs from April 2003 to July 2004 and a newer protocol that PSD officials said they began using in August 2004, after TSA developed a protocol to be used by all the transportation modes.

More specifically, to analyze and categorize specific differences between the two protocols, two analysts compared the first and second protocols to determine the extent to which content from the 73 questions in the newer protocol corresponded with content in the original protocol. To ensure the validity and reliability of our analysis, the two analysts discussed and reconciled any differences. With the assistance of a methodologist, the analysts mutually agreed on how to categorize their assessment of the newer protocol questions. They agreed on the following two categories to describe whether the information could have been reliably transferred from one protocol to the other:

- We were reasonably assured that PSD staff would have been able to accurately transfer completed information from the first protocol to the second.
- We could not be reasonably assured that PSD staff would have been able to accurately transfer completed information from the first protocol to the second.

Because we could not be reasonably assured of the accuracy of the transferred data, we concluded that some of the baseline data key to PSD's outcome measure may not be reliable, as called for in our prior work that describes nine key attributes of successful performance measures.[84] Furthermore, we determined that these data were not sufficiently reliable for the purposes of this report.

We conducted this performance audit from November 2008 to August 2010 in accordance with generally accepted government auditing standards. Those standards require that we plan and perform the audit to obtain sufficient, appropriate evidence to provide a reasonable basis for our findings and conclusions based on our audit objectives. We believe that the evidence obtained provides a reasonable basis for our findings and conclusions based on our audit objectives.

APPENDIX II. COMMENTS FROM THE DEPARTMENT OF HOMELAND SECURITY

U.S. Department of Homeland Security
Washington, DC 20528

Homeland Security

July 23, 2010

Mr. Steve Lord
Director, Homeland Security and Justice Issues
U.S. Government Accountability Office
441 G Street, NW
Washington, DC 20548

Dear Mr. Lord:

Thank you for the opportunity to comment on the draft report titled, *Pipeline Security: TSA Has Taken Actions to Help Strengthen Security, but Could Improve Priority-Setting and Assessment Processes* (GAO-10-867).

The Transportation Security Administration (TSA) values the investigative team's comprehensive review of this Agency's efforts in addressing pipeline security and intends to immediately implement its recommendations. TSA appreciates the professionalism demonstrated by GAO's team members in conducting this difficult and broad-ranging review.

TSA also appreciates GAO's acknowledgment that the Pipeline Security Division (PSD) has (1) identified, consistent with the National Infrastructure Protection Plan, the Nation's most critical pipeline systems; and (2) developed a risk assessment model that combines all three components of risk: threat, vulnerability, and consequence. Further, the GAO report notes that the PSD has made significant progress in completing the requirements of the 9/11 Commission Act, which include establishing a program to review pipeline security plans, initiating inspections of critical facilities of the most essential pipeline systems, developing and promulgating security recommendations, and drafting a Pipeline Security and Incident Recovery Protocol Plan.

The success of TSA's risk-based pipeline security program has been the result of a highly effective public-private partnership and close coordination with other Federal agencies, particularly the U.S. Department of Energy and the U.S. Department of Transportation's Pipeline and Hazardous Materials Safety Administration. As GAO discusses in the report, the PSD has actively engaged with pipeline system operators on a number of security programs. In each of these endeavors, TSA has benefited from the dedicated efforts of the Pipeline Sector Coordinating Council, which plays a significant communications and coordination role between the PSD and the pipeline industry. As an example of this partnership, the International Pipeline Security Fornm, now in its sixth year, is actively supported by industry operator and trade association speakers and attendees. Similarly, pipeline system operators have been enthusiastic participants in TSA's development of training videos in security awareness, improvised explosive device awareness, and pipeline infrastructure security training for law enforcement personnel.

As GAO acknowledges, of particular note is PSD's close coordination with Government and industry partners in developing TSA's Pipeline Security Guidelines. As the guidelines were crafted and refined, the process involved the active participation of pipeline industry representatives in multiple meetings and conference calls. The document, although a voluntary standard, provides TSA's expectations for an effective security program for pipeline operators. The guidelines will serve as the basis for PSD's Corporate Security Reviews and other assessments of the pipeline industry's security status.

The PSD has implemented a risk-based security program that will be enhanced by the adoption of GAO's recommendations. TSA's specific responses to the recommendations are identified below.

Recommendations for Executive Action
To improve aspects of the Pipeline Security Division's (PSD) efforts to help ensure pipeline security, GAO recommends that the Assistant Secretary for the Transportation Security Administration take the following eight actions:

Recommendation 1: To ensure that PSD is managing risk effectively, develop a plan with time frames and milestones for improving the data in the pipeline risk assessment model by, for example, adding more data to the consequence component.

TSA's Response: TSA concurs with this recommendation. The PSD will develop a plan to coordinate the security efforts underway that will help refine the risk ranking tool. Additional data from critical facility inspections, the hazardous liquid pipeline assessment, and toxic inhalation hazard study, among others, will help inform the consequence component.

Recommendation 2: To ensure that PSD is managing risk effectively, document a methodology for scheduling Corporate Security Reviews (CSR) and Critical Facility Inspections (CFIs) that considers a pipeline system's risk ranking as the primary scheduling criteria and balances it with other practical considerations.

TSA's Response: TSA concurs with this recommendation. In TSA's analysis, those pipeline systems that transport the greatest amount of energy were identified as critical. PSD developed the risk ranking tool to further enhance its risk-based effort. To increase the value of this tool in its programs, PSD will develop additional data to inform the tool's rankings and base its programmatic efforts on the results.

Recommendation 3: To help PSD maximize its Corporate Security Review and Critical Facility Inspection efforts and keep its knowledge of the security posture of the pipeline industry current, develop a plan that includes a defined approach and time frame for how and when PSD intends to begin transmitting CSR recommendations in writing to pipeline operators.

TSA's Response: TSA concurs with this recommendation. Although PSD has provided initial briefings at the conclusion of CSRs and subsequently followed up with more extensive briefings by teleconference, the recommendations have not typically been provided in writing. PSD intends to modify this process to ensure that pipeline operators are provided with written recommendations for security improvements.

Recommendation 4: To help PSD maximize its Corporate Security Review and Critical Facility Inspection efforts and keep its knowledge of the security posture of the pipeline industry current, establish a database of CSR recommendations and develop a process for following up on the implementation of those recommendations.

TSA's Response: TSA concurs with this recommendation. PSD will initiate the development of a CSR recommendations database. Repeat reviews of pipeline corporations will particularly focus on the implementation of recommendations from prior reviews.

Recommendation 5: To help PSD maximize its Corporate Security Review and Critical Facility Inspection efforts and keep its knowledge of the security posture of the pipeline industry current, establish a database of CFI recommendations and develop a process for following up on the implementation of those recommendations.

TSA's Response: TSA concurs with this recommendation. PSD has initiated the development of a CFI recommendations database. Following up on these recommendations will enable TSA to assess the pipeline industry's progress in mitigating identified security deficiencies.

Recommendation 6: To better achieve the security strategy laid out in its Pipeline Modal Annex–the national security strategy for pipeline systems–to the extent feasible, revise future updates of the annex to incorporate performance measures for assessing PSD and pipeline industry progress and link those measures to the pipeline security objectives.

TSA's Response: TSA concurs with this recommendation. In future updates to the Transportation Systems Sector Specific Plan PSD will include performance measures within the Pipeline Modal Annex consistent with the sector format and guidance.

Recommendation 7: To better evaluate PSD's performance in helping strengthen the security of hazardous liquid and natural gas pipelines and improvements in pipeline security, develop additional outcome measures that are directly linked to sector goals and modal objectives and track progress towards its stated pipeline security objective.

TSA's Response: TSA concurs with this recommendation. PSD will develop appropriate outcome measures that reflect the impact of its security programs and the security status of the pipeline industry. In so doing, this effort will be made consistent with the performance measurement guidance of the Transportation Systems Sector Specific Plan.

Recommendation 8: To help ensure reliable reporting of security improvements in the pipeline industry, establish reliable baseline data and, until that time, refrain from using reconstructed baseline data to report progress in closing the vulnerability gap.

TSA's Response: TSA concurs with this recommendation. Updated data from repeat Corporate Security Reviews will be utilized to insure more accurate reporting of the pipeline industry's security status.

Thank you for the opportunity to comment on this Draft Report and we look forward to working with you on future homeland security issues.

Sincerely,

Jerald E. Levine
Director
Departmental GAO/OIG Liaison Office

End Notes

[1] Pub. L. No. 107-71, 115 Stat. 597 (2001).

[2] Pub. L. No. 107-296, 116 Stat. 2135 (2002).

[3] Pub. L. No. 110-53, 121 Stat. 266 (2007). The 9/11 Commission was a congressionally chartered commission established by Congress on November 27, 2002, to (1) investigate the relevant facts and circumstances relating to the terrorist attacks of September 11, 2001; (2) identify, review, and evaluate lessons learned from these attacks; and (3) report to the President and the Congress on findings, conclusions, and recommendations that generated from the investigation and review.

[4] The NIPP provides a unifying structure for the integration of a range of efforts for the protection and resilience of the nation's critical infrastructure and key resources.

[5] Throughout this report, we use the term pipelines to refer to either hazardous liquid or natural gas pipelines.

[6] A system is considered critical if it is so vital to the United States that its incapacitation or destruction would have a debilitating effect on security, national economic security, public health or safety, or any combination thereof. PSD determined the most critical pipeline systems based on the amount of energy they carry.

[7] Corporate Security Reviews are on-site reviews to assess corporate security plans for pipeline systems. The intent of these reviews is to develop first-hand knowledge of security planning, establish working relationships with key pipeline security personnel, and identify and share good security practices. PSD has conducted CSRs for the 100 most critical pipeline systems.

[8] PSD established a program for inspecting all the critical facilities of the 100 most critical pipeline systems, as required by the Implementing Recommendations of the 9/11 Commission Act. These physical inspections include the interior and exterior of each critical facility.

[9] The NIPP obligates each sector to develop a sector-specific plan that describes strategies to protect the nation's critical infrastructure and key resources under its purview, outline a coordinated approach to strengthen security efforts, and determine appropriate programmatic funding levels. TSA, as the sector-specific agency for the transportation sector, developed the Transportation Systems Sector-Specific Plan, which describes the strategies to protect all modes of transportation (aviation, maritime, mass transit, highway, freight rail, and pipeline).

[10] The Project Management Institute, *The Standard for Program Management* © (2006).

[11] The 2002 circular outlines voluntary actions that pipeline operators should take and describes actions the federal government plans to take to improve pipeline security. We also reviewed the Pipeline Security Contingency Planning Guidance, which is considered part of the 2002 circular.

[12] Documents we reviewed included PSD's Pipeline Modal Annex, CSR Standard Operating Procedures, CSR and Critical Facility Inspection (CFI) protocols, and Pipeline Security Smart Practices.

[13] Natural Resources Canada is the Canadian government agency that seeks to enhance the responsible development and use of Canada's natural resources and the competitiveness of Canada's natural resources products.

[14] GAO, *Standards for Internal Control in the Federal Government*, GAO/AIMD-00-21.3.1 (Washington, D.C.: November 1999). These standards, issued pursuant to the requirements of the Federal Managers' Financial Integrity Act of 1982 (FMFIA), provide the overall framework for establishing and maintaining internal control in the federal government. Also pursuant to FMFIA, the Office of Management and Budget issued Circular A-123, revised December 21, 2004, to provide the specific requirements for assessing the reporting on internal controls. Internal control standards and the definition of internal control in Circular A-123 are based on Standards for Internal Control in the Federal Government.

[15] Project Management Institute, *A Guide to the Project Management Body of Knowledge* © (Fourth Edition, 2008).

[16] Within TSA, the Office of Transportation Sector Network Management manages all surface transportation security issues with divisions dedicated to each surface mode of transportation, including pipeline.

[17] Exec. Order No. 13,416, 71 Fed. Reg. 71,033 (Dec. 5, 2006). The order mandates that an annex shall be completed for each surface transportation mode in support of the Transportation Systems Sector-Specific Plan.

[18] The NIPP obligates each sector to develop a sector-specific plan that, among other things, describes strategies to protect the nation's critical infrastructure and key resources under its purview. TSA developed the Transportation Systems Sector-Specific Plan, which describes the strategies to protect all modes of transportation, including pipeline.

[19] GAO, *Combating Terrorism: Evaluation of Selected Characteristics in National Strategies Related to Terrorism*, GAO-04-408T (Washington, D.C.: Feb. 3, 2004).

[20] Pub. L. No. 107-71, 115 Stat. 597 (2001).

[21] Pub. L. No. 107-296, 116 Stat. 2135 (2002).

[22] The Pipeline Security Division was established as a separate modal division in November 2005.

[23] API issued a second edition of its guidelines in April 2003 and a third edition in April 2005. INGAA and AGA updated and published their guidelines internally in May 2008.

[24] Although security measures are generally voluntary for operators of critical pipeline facilities, some operators have off-shore or port facilities that are regulated under the Maritime Transportation Security Act and are required to implement certain protective measures.

[25] Pub. L. No. 110-53, §§ 1557, 1558, 121 Stat. 266, 475-77 (2007).

[26] Recognizing that each sector possesses its own unique characteristics and risk landscape, HSPD-7 established sector-specific agencies for each of the critical infrastructure sectors and assigned those agencies responsibility for protecting the critical infrastructure within their area of expertise. HSPD-7 established 17 sectors and DHS later added an 18th sector. The 18 sectors are: agriculture and food; defense industrial base; energy; healthcare and public health; national monuments and icons; banking and finance; water; chemical; commercial facilities; critical manufacturing; dams; emergency services; nuclear reactors, materials, and waste; information technology; communications; postal and shipping; transportation systems; and government facilities. DHS serves as the sector-specific agency for transportation systems and 10 other sectors, and designated TSA as the lead sector-specific agency for transportation, including pipeline.

[27] PSD uses system annual throughput in determining pipeline system criticality, which is based on the amount of hazardous liquid or natural gas product transported through a pipeline in 1 year (i.e., annual throughput). PSD officials told us they purchase a database containing annual pipeline throughput information to determine the 100 most critical pipeline systems and contact pipeline operators to verify information if needed. The 100 most critical systems can shift from year to year. For example, a system might be among the 100 most critical systems one year, but not the next, due to changes that affect each system's throughput. Changes that can affect an operator's position or even presence among the 100 most critical systems include increasing or decreasing annual throughput, going out of business, or selling or purchasing parts or all of a pipeline system.

[28] DHS defines a risk score as a numerical result of a semiquantitative risk assessment methodology and is a numerical representation that gauges the combination of threat, vulnerability, and consequence at a specific moment.

[29] PSD calls its risk assessment model the Pipeline Relative Risk Ranking Tool.

[30] The Office of Intelligence also disseminates additional threat and suspicious incident information related to the pipeline sector to key federal and nonfederal stakeholders, as needed.

[31] "Standard" questions refer to the ones that PSD scores and uses in calculating the CSR score. The CSR protocol includes five additional questions that are not scored, such as questions on the operator's view of the threat to the pipeline industry and how cost affected the operator's ability to implement security enhancements.

[32] GAO, *Highway Infrastructure: Federal Efforts to Strengthen Security Should Be Better Coordinated and Targeted on the Nation's Most Critical Highway Infrastructure*, GAO-09-57 (Washington, D.C.: January 30, 2009); *Commercial Vehicle Security: Risk-Based Approach Needed to Secure the Commercial Vehicle Sector*, GAO-09-85 (Washington, D.C.: February 27, 2009); and *Freight Rail Security: Actions Have Been Taken to Enhance Security, but the Federal Strategy Can Be Strengthened and Security Efforts Better Monitored*, GAO-09-243 (Washington, D.C.: April 21, 2009).

[33] The Project Management Institute, *The Standard for Program Management* © (2006).

[34] DHS's Office of Infrastructure Protection also conducts vulnerability assessments on some pipeline facilities. However, because these assessments are conducted at the facility level rather than the system level, PSD cannot use these assessments in its risk assessment model, which focuses on the system level.

[35] We calculated a simple correlation coefficient to measure the strength and direction of the linear relationship between systems' risk rankings and the time elapsed between PSD's first and subsequent CSRs for the pipeline systems that had two CSRs. This resulted in a correlation coefficient score of 0.2, which indicates a weak correlation. The magnitude of the correlation coefficient determines the strength of the correlation. A perfect correlation equals 1 and no correlation equals 0.

[36] PSD conducts CSRs with operators of the 100 most critical pipeline systems. If an operator owns or operates more than one system among the 100 that are most critical, and uses the same corporate security plan for all its

systems, PSD conducts a single CSR for that operator. As a result, PSD did not need to conduct 100 CSRs to complete CSRs for the 100 most critical pipeline systems.

[37] The line of best fit is found by using the least squares method, which involves finding the minimum of the sum of the squares of the vertical distances of each data point from the proposed line. It is often useful to attempt to represent data with the equation of a straight line in order to predict values that may not be displayed on a scatter plot. The slope of the line of best fit generally does not reflect the magnitude of the correlation.

[38] As of May 2010, PSD had conducted second CSRs for 9 of the top 15 highest risk-ranked systems. PSD conducted first CSRs for the other 6 systems in 2006 or later.

[39] Because some CSRs cover multiple systems (since some operators operate more than one system), we accounted for one system, or one CSR, per operator in our calculations.

[40] We calculated a regression equation to see the extent to which values of the two independent variables—(1) a system's risk ranking and (2) whether PSD had inspected any critical facilities belonging to a system's operator—were associated with values of the dependent variable—i.e., the time elapsed between a first and second CSR. We found little variation in the time elapsed between CSRs that could be explained by the two independent variables. Although PSD officials might have had contact with pipeline operators through other means, we could not quantify other forms of contact and, therefore, could not include them in the analysis.

[41] As of April 2010, 64 systems included in PSD's risk assessment model had at least one critical facility, according to information operators reported, and PSD had inspected at least one critical facility of 43 of these 64 systems. We calculated a simple correlation coefficient to measure the strength and direction of the linear relationship between the systems' risk rankings and when (i.e., the order in which) PSD conducted the first critical facility inspection of that system. This resulted in a correlation coefficient score of 0.03, which indicates almost no correlation.

[42] Although these security measures are generally voluntary for operators of critical pipeline facilities, some operators have off-shore or port facilities that are regulated under the Maritime Transportation Security Act and are required to implement certain protective measures.

[43] If an operator considers none of its facilities to be critical, the operator should document the basis for this conclusion.

[44] INGAA and AGA published security guidelines for the natural gas industry, which were adopted by APGA; API published security guidelines for the petroleum industry.

[45] PSD officials explained they began conducting inspections of critical facilities as part of a new program in November 2008 and curtailed CSR spot checks of selected facilities at that time.

[46] We compared industry guidance to the CSR protocol and found that the protocol generally allows PSD to determine whether a pipeline operator's corporate security plan is consistent with industry guidance.

[47] CSRs include questions pertaining to cyber security, but according to PSD officials, they do not involve in-depth inspections or assessment of an operator's cyber security system and its vulnerabilities because PSD does not possess this expertise. They explained that other federal component agencies, such as DHS's National Cyber Security Division, have this expertise, and pipeline operators typically have in-house expertise or contract for it.

[48] The CSR protocol is divided into the following 11 functional areas: threat assessment, vulnerability assessment, security planning, credentialing, secure areas, critical infrastructure, physical security, cyber security, security training, communications, and exercises.

[49] The 9/11 Commission Act requires DHS to establish a program for reviewing pipeline operator adoption of the recommendations of the 2002 circular, including the review of pipeline security plans, and requires DHS to develop and implement a plan to review the pipeline security plans of the 100 most critical pipeline operators covered by the 2002 circular. Pub. L. No. 110-53, § 1557(a), (b), 121 Stat. 266, 475 (2007).

[50] Because PSD updates the 100 most critical systems annually using pipeline system energy throughput data, which is revised annually, PSD has conducted CSRs of operators whose systems once were, but may no longer be, on the most critical list. Also, as noted earlier, because some pipeline operators own or operate more than one of the 100 most critical systems, PSD did not need to conduct 100 CSRs to cover all 100 most critical systems.

[51] The 9/11 Commission Act requires DHS to establish a program for reviewing pipeline operator adoption of the recommendations of the 2002 circular, including critical facility inspections, and requires DHS to develop and implement a plan to inspect the critical facilities of the 100 most critical pipeline operators covered by the 2002 circular. Pub. L. No. 110-53, § 1557(a), (b), 121 Stat. 266, 475 (2007).

[52] Toxic inhalation hazard pipelines, such as those transporting anhydrous ammonia and chlorine gas, are among the most dangerous. These pipelines, which have relatively low energy throughputs, are not addressed by the 2002 circular or the 9/11 Commission Act; nevertheless, PSD officials have told us the security of these pipelines is important and should be addressed.

[53] In trying to recall the origin of the decision to not communicate recommendations in writing, PSD officials said it was based on concerns about an operator's potential liability if it did not implement the recommendations and its pipeline system was later attacked. However, officials acknowledged that they send operators written recommendations for their newer program—the CFI program—without such concerns.

[54] PSD has contracted with Johns Hopkins University Applied Physics Laboratory to revise the CSR protocol.

[55] Project Management Institute, *A Guide to the Project Management Body of Knowledge* © (Fourth Edition, 2008).

[56] During the course of our review, the number of PSD staff ranged from 11 to 12. Three of these staff generally conducted CSRs.

[57] Some pipelines may be vulnerable to "cyber attacks" on computer control systems that are used to collect data from pipeline sensors in real time and display these data to controllers, who monitor the data and operate pipeline control equipment remotely. A pipeline operator's control system represents a significant investment on the part of the operator and is a critical resource for response and recovery in the event of a pipeline incident of almost any type.

[58] The 9/11 Commission Act states that interstate and intrastate transmission and distribution pipeline operators, nonprofit employee organizations representing pipeline employees, emergency responders, offerors, state pipeline safety agencies, public safety officials, and any other relevant parties are to be consulted. The incident recovery protocols plan is also to be developed in conjunction with interstate and intrastate pipeline operators and terminal and facility operators connected to pipelines.

[59] The Office of Infrastructure Protection leads the coordinated national program to reduce risks to the nation's critical infrastructure and key resources posed by acts of terrorism, and to strengthen national preparedness, timely response, and rapid recovery in the event of an attack, natural disaster, or other emergency.

[60] PSD officials reported to us that they had coordinated the plan with DHS/TSA components and other DHS components, DOT/PHMSA, DOE, Department of Justice/FBI, Department of Interior/Minerals Management Service, National Transportation Safety Board, Federal Energy Regulatory Commission, Environmental Protection Agency, Federal Energy Regulatory Commission, Department of Defense/U.S. Army Corps of Engineers, National Association of Regulatory Utility Commissioners, National Association of State Energy Officials, National Governors Association, National Emergency Managers Association, National Association of Pipeline Safety Representatives, International Association of Fire Chiefs, International Association of Chiefs of Police, National Sheriff's Association, Pipeliners Union Local 798, Interstate Natural Gas Association of America, and Association of Oil Pipe Lines Owners/Operators.

[61] In prior work we identified a set of desirable characteristics to aid responsible parties in further developing and implementing national strategies, and to enhance the usefulness of those strategies in resource and policy decisions and better ensure accountability. For a more detailed discussion of these characteristics, see GAO, *Combating Terrorism: Evaluation of Selected Characteristics in National Strategies Related to Terrorism*, GAO-04-408T (Washington, D.C.: Feb. 3, 2004).

[62] The Pipeline Modal Annex also identifies supporting strategies PSD will pursue to achieve pipeline security objectives and presents information to explain what TSA, other federal components, or industry is doing and how those activities correspond with these strategies. For example, the Pipeline Modal Annex describes the CSR program as a program to promote the implementation of layered threat deterrence and vulnerability mitigation programs and to conduct network enhancement and information-sharing activities.

[63] According to PSD officials, they have prepared a 2010 revision to the 2007 Pipeline Modal Annex, which also does not incorporate performance measures and milestones. Officials told us in May 2010 that the revised annex was in internal review.

[64] See GAO-04-408T.

[65] Each CFI trip involves inspections of multiple critical facilities.

[66] According to PSD and TSNM officials, an appendix to the 2010 Sector Critical Infrastructure and Key Resources Protection Annual Report for the Transportation Systems Sector will discuss other performance measures related to two risk mitigation activities—(1) the percentage of the 100 most critical pipeline systems that have had a CSR or a repeat CSR and (2) the percentage of the 100 most critical systems that have conducted annual security exercises and drills (specifically, the percentage that has participated in Intermodal Security Training Exercise Program exercises). As of May 2010, this report was in internal review.

[67] TSA's Intermodal Security Training Exercise Program offers an intermodal transportation security exercise program for transportation sector network communities. The program is intended to enhance the preparedness of the nation's surface transportation sector network with evaluations of prevention, preparedness, and the ability to respond to terrorist-related incidents.

[68] GAO, *Risk Management: Further Refinements Needed to Assess Risks and Prioritize Protective Measures at Ports and Other Infrastructure*, GAO-06-91 (Washington, D.C.: December 15, 2005.)

[69] The Pipeline Modal Annex identifies other programs and activities that seek to increase resiliency and robustness.

[70] GAO, *Aviation Security: A National Strategy and Other Actions Would Strengthen TSA's Efforts to Secure Commercial Airport Perimeters and Access Controls*, GAO-09-399 (Washington, D.C.: September 30, 2009).

[71] PSD subsequently made minor revisions to the second CSR protocol that did not affect our analysis or the data PSD uses for its outcome measure.

[72] According to PSD officials, they completed 31 CSRs from mid-April 2003 through mid-July 2004.

[73] GAO, *Tax Administration: IRS Needs to Further Refine Its Tax Filing Season Performance Measures*, GAO-03-143 (Washington, D.C.: Nov. 22, 2002). In this report, GAO reported on nine key attributes of successful performance measures including the reliability of measures.

[74] GAO, *Managing for Results: Challenges Agencies Face in Producing Credible Performance Information,* GAO-GGD-00-52 (Washington, D.C.: Feb. 4, 2000).
[75] An operator's CSR score is calculated based on the 73 standard questions in the newer CSR protocol.
[76] For pipeline operators that operate more than one system, we used only the highest risk-ranked system for that operator in our analysis to control for the possibility that PSD also conducted a second CSR for a lower risk system belonging to the same operator.
[77] Although PSD officials might have contact with pipeline operators through means other than CSRs and CFIs, we could not quantify other forms of contact and, therefore, could not include them in the analysis.
[78] The Project Management Institute, *The Standard for Program Management* © (2006).
[79] Our review of the 2002 circular included the Pipeline Security Contingency Planning Guidance.
[80] GAO, *Standards for Internal Control in the Federal Government,* GAO/AIMD-00-21.3.1 (Washington, D.C.: November 1999).
[81] Project Management Institute, *A Guide to the Project Management Body of Knowledge* © (Fourth Edition, 2008).
[82] Exec. Order No. 13,416, 71 Fed. Reg. 71,033 (Dec. 5, 2006).
[83] GAO, *Combating Terrorism: Evaluation of Selected Characteristics in National Strategies Related to Terrorism,* GAO-04-408T (Washington, D.C.: Feb. 3, 2004).
[84] GAO, *Tax Administration: IRS Needs to Further Refine Its Tax Filing Season Performance Measures*, GAO-03-143 (Washington, D.C.: Nov. 22, 2002).

In: Pipeline Safety and Security
Editor: Jonathan T. Miller and Lisa M. Gurton

ISBN: 978-1-61209-147-1

Chapter 3

"THE SAFETY OF HAZARDOUS LIQUID PIPELINES (PART 2): INTEGRITY MANAGEMENT" HEARING, BEFORE THE SUBCOMMITTEE ON RAILROADS, PIPELINES, AND HAZARDOUS MATERIALS

Jonathan T. Miller and Lisa M. Gurton

PURPOSE OF THE HEARING

The Subcommittee on Railroads, Pipelines, and Hazardous Materials is scheduled to meet on Thursday, July 15, 2010, at 10:00 a.m., in room 2167 of the Rayburn House Office Building to receive testimony on pipeline operators' management of the safety of hazardous liquid pipelines, known as integrity management. The hearing is the third in a series of hearings that the Subcommittee will conduct on pipeline safety.

BACKGROUND

According to the Pipeline and Hazardous Material Safety Administration (PHMSA), there are more than 170,000 miles of onshore and offshore hazardous liquid pipelines (about 200 operators) in the United States, which carry more than 75 percent of the nation's crude oil and around 66 percent of its refined petroleum products. Of the more than 170,000 miles of hazardous liquid pipeline, about 55,000 miles are major crude oil trunk lines, which range in diameter from eight inches to 48 inches. Associated with these trunk lines in several locations is significant crude oil tankage, and about 30,000 to 40,000 miles of crude gathering lines, which are smaller lines that gather the oil, gas, and water from many wells, both onshore and offshore, and connect to the larger trunk lines. In addition, there are about 95,000 petroleum product lines, flow lines/piping associated with well operations, and produced water pipelines (containing contaminated water following oil, gas, and water separation). Many hazardous liquid pipelines also transport highly volatile liquid, which is hazardous liquid that will form a vapor cloud when released to the atmosphere.

According to PHMSA, on average over the last five years, there were 349 reportable hazardous liquid pipeline incidents, resulting in two fatalities and four injuries requiring in-patient hospitalization and $131,900,023 in property damage. Although pipeline releases have caused relatively few fatalities in absolute numbers, a single pipeline incident can be catastrophic.

I. Managing the Structural Integrity of Hazardous Liquid Pipelines

On February 1, 2000, in the wake of several pipeline ruptures in Bellingham, Washington, Simpsonville, South Carolina, Reston, Virginia, and Edison, New Jersey, PHMSA issued a final rule requiring pipeline operators to evaluate the potential consequences of failure of their pipeline segments that could affect a high consequence area (HCA),[1] and set priorities for inspecting, operating, and maintaining the pipeline based on whether people, property, or the environment might be at risk should a pipeline failure occur.[2] According to PHMSA, pipeline segments that could affect an HCA represent about 44 percent of the total hazardous liquid pipeline mileage in the United States.

Specifically, all hazardous liquid pipelines operators are required to determine which of their pipeline segments could affect HCAs on an ongoing basis. For example, an area that an operator determined did not affect an HCA might subsequently affect an HCA, depending on the circumstances (e.g., new high population or environmentally sensitive areas). After the pipeline segments are identified, the operator is required to comprehensively assess the structural integrity of those pipeline segments that could affect HCAs, using a variety of assessment methods determined appropriate by the operator. Based on these assessments, operators must take prompt action to repair any defects that could reduce a pipeline's integrity. Integrity management assessments must be performed at least once every five years. However, an additional eight months may be added to the reassessment interval to allow for unforeseeable events (e.g., permitting delays, weather events, tool failures) that could affect the ability of the operator to successfully complete an assessment.

As part of the program, PHMSA requires each hazardous liquid pipeline operator to maintain a written integrity management plan at the operators' place of business. Federal regulations specify eight elements that a pipeline operator's integrity management plan must include, but the plan is not submitted to PHMSA for review or approval prior to implementation of the plan (or when a plan is revised).[3] The plans are provided to PHMSA (at PHMSA's request) just prior to each Federal inspection, but they are not maintained by PHMSA because the plans frequently change. According to PHMSA, since 2002, each pipeline operator has been inspected at least one time to review its integrity management program and each major operator has been inspected twice.

In addition to maintaining a written integrity management plan, each operator is required to submit to PHMSA annual performance measure reports,[4] which provide the total miles of each operator's pipelines; the total miles that could affect HCAs; the total volume of product transported in barrel-miles; the total mileage of pipeline assessed by the operator; the total number of anomalies identified as a result of those assessments; and the total number of conditions repaired.[5] Although the reports are publicly available, they provide no information to PHMSA and the public on the actual nature of the anomalies identified by the pipeline operator during the inspections and the conditions that required repair. Safety advocates maintain that more detailed reporting could provide valuable information to PHMSA and the public about the condition of the pipeline and the operating and maintenance practices of the operator.

The first round of operator-performed assessments was completed in February 2009. Pipeline operators reported to PHMSA that they made 31,855 repairs to hazardous liquid pipeline segments that, if left unaddressed, could have affected HCAs. Of those, 6,831 defects were considered to be so serious that immediate repair was required under the regulations; another 25,024 hazardous liquid defects had to be repaired within a 60- to 180-day time period.[6] An example of an immediate repair would be wall loss of more than 80 percent. Certain dents (by size) must be repaired within 60 and 180 days, and 50 percent or more wall loss must be repaired within 180 days.

Safety advocates maintain – particularly given the large number of defects identified as a result of the assessments – that PHMSA should expand the scope of the integrity assessments to require pipeline operators to evaluate the integrity of their pipelines outside of HCAs. Pipeline operators, however, maintain that while they are only required to assess pipelines that could affect HCAs, in practice they evaluate a much greater percentage of pipelines when they conduct these assessments. According to the operators, this is due largely to the practical constraints associated with running in-line inspection tools, such as smart pigs; because of the location of the launchers and receivers used to insert and remove smart pigs from the pipeline, relatively long sections of pipeline are inspected when these tools are used. These sections generally contain portions of the pipeline that can affect HCAs and portions of the pipeline that do not affect HCAs. Thus, while conducting assessments of the portions of their lines that affect HCAs, operators running smart pigs also obtain data on the condition of their pipelines in other areas and take action to assure the integrity of those sections outside of HCAs.

Safety advocates, however, maintain that although larger areas may be assessed, Federal regulations only require operators to repair and report any defects identified that could affect HCAs. If an operator identifies a defect outside of an HCA as a result of those larger assessments, that operator is not required to repair that defect or report it to PHMSA under existing regulations.

II. How Operators Assess the Integrity of Their Pipelines

Federal regulations allow pipeline operators to determine the best method(s) of assessing the structural integrity of their pipelines, using one or more of the following three approaches: in-line inspection (ILI), hydrostatic testing, or direct assessment.[7] Alternative assessment methods can be employed if they can be shown to be effective.[8]

ILI, also known as "pigging", is used to detect wall thickness and the amount of corrosion in the line providing the operator with information on operability and safety. Pigs have been an integral part of maintaining pipelines since the beginning of the 20th Century. The earliest devices were basic utility pigs, better known as scraper pigs. Updated versions are still in use today, scraping and scrubbing pipes to remove liquid and solid buildup.[9]

Since 1965, oil pipeline operators have used technologically advanced versions that measure and record problems in the pipes. Known as "smart pigs", these mechanical devices check for potential problems such as corrosion, dents, and cracks, and provide information to a pipeline operator so that corrective measures can be taken.

Smart pigs are cylinder-shaped electronic devices inserted into the pipe and then propelled by the flowing oil. They detect loss of metal or deformations in the pipeline. Pipeline operators use this data to determine where potential problems are, which are then investigated further and repaired as needed.

New technologies have made smart pigs even more efficient and effective. Pigs that use magnetic flux leakage (MFL) technology, ultrasonic measurements, and geometric tools are among the most common smart pigs used today.

There are two types of MFL technologies in use: standard and transverse flux inspection (TFI). TFI operates the same way as standard MFL, except that the magnetic field it generates is turned 90 degrees. Standard MFL pigs are best at detecting cracks and other defects while TFI pigs are better at detecting seam-related corrosion.

MFL can detect corrosion by sensing magnetic leakage. First, it initiates a magnetic field in the pipeline. If there are any flaws in the pipeline wall, some of the magnetic field will escape. Sensors onboard the pigs detect and measure that leakage. Smart pigs equipped with MFL technology can determine whether the corrosion is internal or external, and they can also measure for changes in the thickness of the walls.

Measurements are made when the pigs emit ultrasonic signals whose echoes are timed and compared with data to determine the wall's thickness. The same ultrasonic technology can detect longitudinal cracks, crack-like defects, and longitudinal weld defects.

Smart pigs can also identify deformations, dents, or obstructions by measuring the bore of the pipe for uniformity. These tools utilize mechanical arms or electro-mechanical instruments.

Pigging has the potential to provide for 100 percent coverage, in contrast to other assessment methods. However, certain larger defects may be missed and defects may not be sized correctly. This is why it is important that each pipeline operator utilize the best assessment tool, and that may require more than one tool.

Hydrostatic testing involves filling a section of pipe with water and increasing the pressure to a level significantly above the normal operating pressure.[10] The primary purpose of hydrostatic testing is to detect and remove joints of the pipeline that contain defects (including corrosion pits or cracks) by causing them to leak or rupture while the pipeline is filled with water.[11]

Direct assessment involves obtaining information from existing records on pipelines, taking measurements of the pipeline, excavating and examining the pipe, and analyzing post-assessment data.[12] Direct assessment is often used for unpiggable pipelines where an interruption of service would be impractical.[13]

III. Background on Alyeska Pipeline Company

In 1968, oil was discovered at Prudhoe Bay in the North Slope, located in northern Alaska between the Brooks Range Mountains and the Beaufort Sea (part of the Arctic Ocean). A consortium of oil companies planning to produce the oil determined that a pipeline offered the best means to transport crude oil from the North Slope to a navigable port in southern Alaska where it could be shipped by tanker to refineries in the continental United States. The pipeline route would cover 800 miles from Prudhoe Bay to the port of Valdez, the northernmost ice-free port in the United States.[14]

The Alyeska Pipeline Service Company was established in 1970 and charged with designing, constructing, operating, and maintaining the Trans Alaska Pipeline System, commonly called TAPS. Pipeline construction began in March 1975 and was finished in June 1977. Crude oil began flowing in the pipeline on June 20, 1977, and the first tanker filled with North Slope crude oil left Valdez on August 1, 1977.[15]

The 48-inch diameter steel pipeline runs 800 miles and crosses three mountain ranges and over 800 rivers or streams; 420 miles of the pipeline is elevated above ground to keep the permafrost from melting. The volume of oil flowing through the pipeline has decreased from a peak of 2.1 million barrels per day (bpd) in 1988 to about 650,000 bpd in 2010. Alaska today supplies nearly 14 percent of the United States' domestic crude oil production.

The consortium of companies that own TAPS today includes:

BP Pipelines (Alaska) Inc.:	46.93 percent
ConocoPhillips Transportation Alaska, Inc.:	28.29 percent
ExxonMobil Pipeline Company:	20.34 percent
Koch Alaska Pipeline Company, LLC:	3.08 percent
Unocal Pipeline Company:	1.36 percent[16]

BP has a strong role in directing Alyeska's operations. Alyeska's president, Kevin Hostler, spent 27 years with BP, most recently as senior vice president of BP's global human resources organization. Before that, Hostler was head of BP's subsidiary in Columbia. According to recent press reports, a top BP Alaska official asked, in light of the Gulf disaster, whether it is a good idea to have Mr. Hostler, a "BP executive", running TAPS, "where BP can exert cultural and economic influence through the president of [Alyeska] as well as its ownership share, in directions that are not good for the safety and the integrity of [the pipeline]." The BP Alaska official said the fact that both companies are plagued by the same safety and management concerns is evidence of a "pervasiveness of a BP leadership culture that is focused on cost cutting that reduces operational integrity." Last Wednesday, one day after the press report surfaced, Alyeska's president resigned, amid allegations that Alyeska – at the direction of BP and Mr. Hostler – was making dangerous cost cuts to the pipeline's integrity management program and falsely inflating the amount of money Alyeska spent on pipeline corrosion.[17] According to Alyeska, Mr. Hostler had already planned to retire at the end of this year.

PHMSA has also been concerned about corrosion in Alyeska's pipelines. In March 2006, internal corrosion on a 34-inch low-stress pipeline, owned by BP Exploration, caused a 5,000 barrel crude oil spill (212,252 gallons spilled) on the North Slope of Alaska. The oil spill was the worst in the history of oil development on Alaska's North Slope, and went undetected for five days before a BP oilfield worker detected the scent of hydrocarbons during a drive through the area. A few months later, in August 2006, a second leak was discovered while BP was inspecting the Eastern Operating Area segment of the pipeline. Field inspection of the leak site revealed multiple holes at a single location, contributing to an estimated spill of about 1,000 gallons of processed crude oil.

During the course of its investigations of the 2006 BP pipeline failures, PHMSA became concerned about the safety of another pipeline (Affected Pipeline) operated by BP and Alyeska because of its shared operating and internal characteristics to the failed BP lines and because the Affected Pipeline had not been cleaned or internally inspected. According to BP, the pipeline is "one of the most important pipelines on the North Slope. The entire flow from the Prudhoe Bay Unit is transported through this pipeline just before it enters the Trans Alaska Pipeline."[18]

Since 2006, PHMSA has – on numerous occasions—urged both BP and Alyeska to comprehensively assess the structural integrity of the Affected Pipeline, repair any defects that were identified, and conduct other work to correct significant corrosion problems or fully replace the pipeline. According to PHMSA, these measures were needed "to protect life, property, and the environment from potential hazards associated with the Affected Pipeline." BP and Alyeska were initially unresponsive to PHMSA's concerns. As a result, on June 20, 2008, PHMSA issued a corrective action order to BP and Alyeska directing them to take immediate action.[19] According to PHMSA, BP and Alyeska now plan to replace the pipeline by the end of this year.

More recently, PHMSA has been focused on a May 25, 2010, incident at Alyeska's Pump Station 9, located near Delta Junction, Alaska. On the morning of May 25, Alyeska briefly shut down the Trans Alaska Pipeline to perform routine maintenance and testing on the fire control system at Pump Station 9 (the site of several previous maintenance failures, including a 2007 fire). Similar maintenance and testing activities had been completed during pipeline slowdowns at other pump stations in early May. During the third and final fire system test procedure at Pump Station 9, which simulated an electrical fire, the station lost power. The redundant power supplies also failed. As a result, Alyeska's operations control center in Anchorage lost visibility to and control of Pump Station 9; meaning, the control center could not see what was happening at the pump station. According to Alyeska's internal investigation report, the station was designed so that a loss of power would cause the station's relief valves to open allowing crude oil to flow into relief tanks so that the pipeline is not inadvertently over pressured. Because of that, crude oil began pumping from the pipeline into an overflow tank, Tank 190. Tank 190 is able to hold 55,000 to 60,000 barrels of oil; prior to the release, it had about 22,000 barrels in it. For a yet-to-be-determined length of time after the pump station lost power, the crude oil filled Tank 190 and then spilled over the top. It was not until the oil spilled over the top that Alyeska personnel who were performing the fire testing observed that crude oil was discharging from the vents on Tank 190 onto the ground.[20] PHMSA estimates that from the time the pump station went "blind" to the control center in Anchorage, about 33,000 barrels (1,386,000 gallons) of oil had filled the tank and another 5,000 barrels (210,000 gallons) spilled onto the ground.

According to Alyeska, a number of significant incidents on TAPS over the last several years, demonstrate a trend of operational discipline deficiencies similar to those involved with the Tank 190 overfill incident.[21] The report also notes that while the operations command center did not have visibility of Pump Station 9 during the overflow incident, monitoring upstream pressure at Pump Station 8 would have provided insight to the controllers that pipeline pressure was dropping, indicating flow to Pump Station 9.

After evaluating the facts surrounding the incident, PHMSA's Associate Administrator for Pipeline Safety issued a corrective action order to Alyeska stating, "I find that the operation of TAPS [the Trans Alaska Pipeline System] without corrective action measures would be hazardous to life, property and the environment." According to PHMSA:

> Additionally, after considering the age of the pipeline facility, the particular circumstances surrounding this failure and crude oil spill, the failure to actuate certain valves around the time of the spill, suspected problems with the electrical power system, the proximity of the pipeline and breakout tank to an HCA, the hazardous nature of the crude oil being transported, the pressure required for transporting the material when the line is operational, Alyeska's intention to restart the pipeline without the relief capacity provided by Tank 190, the uncertainties as to the cause of the failure, and the ongoing investigation to determine the cause of the failure, I find that a failure to issue this order expeditiously to require immediate corrective action would result in likely serious harm to life, property, and the environment.[22]

The corrective action order required Alyeska to implement a variety of safety measures, including locating personnel at the pump station around the clock to monitor activities.[23] Alyeska is in the process of removing personnel from its pump stations along the 800-mile pipeline, as part of its Strategic Reconfiguration plan that, according to Alyeska, concentrates on reducing physical infrastructure and simplifying operations and maintenance. The plan calls for electrification of pump stations and installation of new control systems for the pipeline. After reconfiguration, each pump station will be manned from Alyeska's operations control center in Anchorage. According to Alyeska's prior president, David Wight, who also served as President of BP Amoco Energy Company: "When we are done, Strategic Reconfiguration will shave millions of dollars from the annual cost of moving oil from Prudhoe Bay to Valdez. Our workers still face many challenges as we modernize the pipeline. However, I am confident that they will succeed in extending TAPS' economic life while maintaining the highest safety, integrity and environmental standards."

In addition to the two corrective action orders, PHMSA has issued 13 notices of probable violation to Alyeska since 2002 alleging violation of various Federal regulations, including significant deficiencies in Alyeska's integrity management and corrosion control programs; 10 notices of amendment, which identify shortcomings in Alyeska's plans and procedures under PHMSA regulations; and 10 warning letters regarding deficiencies in Alyeska's programs. PHMSA has proposed a total of $1,754,300 in civil penalties as a result of the alleged failures.

One of the most recent violations was for what PHMSA characterizes as a "near miss incident" in January 2010. BP was using natural gas to push a cleaning pig through its pipeline (an abnormal and potentially dangerous operation). Two breakout tanks at Alyeska's Pump Station 1 were over pressurized due to the rapid influx of natural gas into TAPS, which caused Alyeska's tanks' relief vents and "blow out" type hatches to open and release flammable vapors. Following the incident, Alyeska restarted the pipeline without verifying and confirming system integrity. According to the corrective action order, Alyeska didn't even conduct a visual inspection of the tanks before restarting the system.[24] Alyeska is challenging the proposed violation.

IV. Background on Enbridge – U.S. Operations

Enbridge Pipelines Inc. operates the world's longest and most sophisticated crude oil and petroleum products pipeline system. The 1,900-mile Lakehead System (the U.S. portion of the world's longest pipeline) has operating for 59 years and is the primary transporter of crude oil from Western Canada to the United States. The system spans from the international border near Neche, North Dakota, to the international border near Marysville, Michigan, with an extension across the Niagara River into the Buffalo, New York area. It consists of approximately 3,500 miles of pipe with diameters ranging from 12 to 48 inches; 60 pump station locations; and 64 crude oil storage tanks with a capacity of about 11.6 million barrels. Total deliveries on the Lakehead System averaged 1.62 million bpd in 2008, meeting approximately 72 percent of Minnesota refinery capacity; 64 percent of the greater Chicago area; and 68 percent of Ontario's refinery demand.

Enbridge's North Dakota System is a 330-mile crude oil gathering and 620-mile interstate transportation system that gathers crude oil from points near producing wells in 22 oil fields in North Dakota and Montana. Most deliveries from the approximately 161,000 bpd North Dakota System are made at Clearbrook, Minnesota, which provides connections with the Lakehead System and a third-party pipeline that transports crude oil to refineries in the Minneapolis/St. Paul area. In response to increasing crude oil production in the region, Enbridge recently completed a 51,000 bpd expansion in early 2010. Enbridge is proposing additional expansions of the North Dakota System.

The Enbridge Toledo Pipeline connects to the Lakehead System at Stockbridge, Michigan, and travels southward to two refineries in the Toledo, Ohio, area. This 35-mile pipeline has a capacity of 100,000 barrels per day in heavy crude oil service and became available for service in February 1999.

The first continuous delivery of western Canadian crude oil was initiated in early March 2006 through Enbridge's Spearhead Pipeline to Cushing, Oklahoma. The 650-mile, 22- and 24-inch diameter pipeline historically operated in south-to-north service, but Enbridge bought the pipeline and reversed its flow to provide Canadian crude oil producers and shippers with access to markets in the Mid-Continent and southern United States. Enbridge recently expanded the Spearhead Pipeline, increasing the average annual capacity from 125,000 bpd to 190,000 bpd.

The Mid-Continent System comprises the Ozark and West Tulsa pipelines and storage terminals at Cushing, Oklahoma, and El Dorado, Kansas. The Mid-Continent System includes more than 480 miles of crude oil pipelines, with average deliveries in 2008 totaling 231,000 bpd. It has 96 individual storage tanks ranging in size from 55,000 to 575,000 barrels. A recent expansion of the Cushing Terminal increased storage capacity on the system to 16 million barrels.

In addition to Enbridge's many wholly owned pipelines and facilities, the company has interest in several other liquids pipelines across the United States.[25]

PHMSA reports that, since 2002, it has issued one corrective action letter and one proposed corrective action letter to Enbridge. The most recent corrective action letters stems from a pipeline rupture that occurred on January 8, 2010, resulting in the release of about 3,000 barrels (126,000 gallons spilled) of crude oil near Neche, North Dakota. PHMSA found that continued operation of the line without corrective action measures, including a mandatory reduction in pressure on the pipeline stemming from the Canadian border to Superior, Wisconsin, would be hazardous to life, property, and the environment.

PHMSA proposed another corrective action order (and $2,405,000 in civil penalties) to Enbridge stemming from a fatal accident that occurred on a 34-inch crude oil pipeline on November 28, 2007, near Clearbrook, Minnesota.[26] The oil spill started with a leak the size of a pinhole, and erupted into an explosion and towering flames, killing two Enbridge workers and resulting in more than $2 million in property damage.

PHMSA found that Enbridge failed to: (1) follow its written procedures regarding welds+ends couplings used by its field personnel for pipeline replacement; (2) ensure that the pipeline was sufficiently anchored before installing the couplings; review the work performed by its personnel; (3) remove multiple ignition sources around the area the work was being performed; and (4) ensure that personnel received proper training and were qualified to conduct the repairs. PHMSA also found that Enbridge had operated the pipeline in excess of the design pressure of the weld+ ends couplings as determined by the manufacturer and outlined in its recommended installation instructions.[27]

In addition to the $2,405,000 in civil penalties proposed as a result of the Clearbrook, Minnesota incident, PHMSA has proposed $121,500 in civil penalties for various violations since 2002, including the failure of Enbridge to conduct certain inspections and maintain certain records. PHMSA has also issued six notices of amendment, which identify shortcomings in Enbridge's plans and procedures under PHMSA regulations; and seven warning letters regarding deficiencies in Enbridge's programs.[28]

End Notes

[1] HCAs include unusually sensitive environmental areas (defined in 49 C.F.R. 195.6), urbanized areas, and other populated places as delineated by the United States Census Bureau, and commercially navigable waterways.

[2] *See* PHMSA, *Frequently Asked Questions*, http://www.phmsa.dot.gov/about/faq.

[3] 49 C.F.R. § 195.452.

[4] U.S. Department of Transportation, *Implementing Integrity Management for Hazardous Liquid Operators Performance Measure Reports and Quick Facts*, http://primis.phmsa.dot.gov/iim/perfmeasures.htm.

[5] *Id.*

[6] PHMSA, Power Point presentation, *The Pipeline Inspection Program*, prepared upon request of House Transportation and Infrastructure Committee Majority Staff (March 2010).

[7] Michael Baker Jr., Inc., *Pipeline Corrosion Final Report*, U.S. Department of Transportation Pipeline and Hazardous Materials Safety Administration Office of Pipeline Safety (November 2008), at 33, http://primis.phmsa.dot.gov/iim/docstr/FinalReport PipelineCorrosion.pdf.

[8] *Id.* Operators must advise PHMSA 90 days in advance of conducting an assessment with an alternative technology. *See also* PHMSA, *Implementing Integrity Management for Hazardous Liquid Operators: Frequently Asked Questions* (October 15, 2009), at 17, http://primis.phmsa.dot.gov/iim/faqs.htm.

[9] "Utilizing a Smarter Pig," *In the Pipe: News from the Pipeline Industry* (February 2006).

[10] *Id.* at 35.

[11] *Id.*

[12] *Id.* at 37.

[13] *Id.* at 36.

[14] *See* Alyeska's website at http://www.alyeska-pipe.com.

[15] *Id.*

[16] *Id.*

[17] Jason Leopold, "Dangerous Cost Cuts at Alyeska Pipeline: Yet Another Example of How BP Runs Things," *Truthout*, http://www.truth-out.org/alyeska-pipeline-yet-another-example-how-bp-runs-things61097. *See also* Jason Leopold "Alyeska CEO Steps Down Following Truthout Exposé," *Truthout*, http://www.truth-out.org/alyeska-ceo-resignsfollowing-truthout-expose61 134.

[18] *See* http://primis.phmsa.dot.gov/comm/reports/enforce/Actions opid 261 49.html.

[19] *Id.*

[21] These significant incidents include a tank vent fire at Pump Station 9 on January 6, 2007; a Remote Gate Valve leak on January 9, 2007; an overfill near loss of Tank 190 at Pump Station 9 on March 22, 2007; energy isolation near loss events at Pump Station 9 in October 2008; Sadelrochit Stream Gas Excursion at Pump Station 1 on January 15, 2009; and a pipeline overpressure event at Pump Station 9 on July 19, 2009. *See also* Alyeska Pipeline Service Company, *TK190 Overfill Incident Root Cause Analysis Report and Post Accident Review* (June 22, 2010).

[22] *Id.*

[23] *Id.*

[24] *Id.*

[25] *See* Enbridge Pipeline's Internet web site at http://www.enbridgeus.com/.

[26] *See* http://primis.phmsa.dot.gov/comm/reports/enforce/Actionsopid26149.html.

[27] *See* http://primis.phmsa.dot.gov/comm/reports/enforce/Actions opid 261 49.html.

[28] *Id.*

In: Pipeline Safety and Security
Editor: Jonathan T. Miller and Lisa M. Gurton
ISBN: 978-1-61209-147-1

Chapter 4

WRITTEN STATEMENT OF CYNTHIA L. QUARTERMAN, ADMINISTRATOR, PIPELINE AND HAZARDOUS MATERIALS SAFETY ADMINISTRATION, U.S. DEPT. OF TRANSPORTATION, BEFORE THE SUBCOMMITTEE ON RAILROADS, PIPELINES, AND HAZARDOUS MATERIALS, HEARING ON "PREVENTING SPILLS FROM OIL PIPELINES THROUGH INTEGRITY MANAGEMENT; LEAK DETECTION; SHUT-OFF VALVES; AND CORROSION PREVENTION"

Cynthia L. Quarterman

Chairman Brown, Ranking Member Shuster, members of the Subcommittee, thank you for the opportunity to appear today. Secretary LaHood, the employees of the Pipeline and Hazardous Materials Safety Administration (PHMSA), and the entire Department of Transportation (DOT) share public safety as their top priority. The Department holds a strong commitment to preventing spills on all pipelines through aggressive regulation and oversight to ensure the safety and reliability of the nation's pipeline transportation infrastructure.

PHMSA aims to improve the integrity of pipeline systems and reduce risks. This is fundamentally why PHMSA exists. In virtually every decision PHMSA makes it asks: *how does this help reduce the risk of a pipeline system failure?* Historically, PHMSA regulations consisted of prescriptive measures pipeline operators need to follow. However, over time it became clear that to adequately evaluate risk, it was necessary to understand all the system implications. To that end, the Hazardous Liquid Pipeline Integrity Management Program was created with the following objectives:

- ensuring the quality of pipeline integrity in areas with the highest potential for adverse consequences (high consequence areas or HCAs);
- promoting a more rigorous and systematic management of pipeline integrity and risk by operators;
- maintaining the government's prominent role in the oversight of pipeline operator integrity plans and programs; and
- increasing the public's confidence in the safe operation of the nation's pipeline network.

PHMSA's older prescriptive regulations and newer pipeline Integrity Management Programs work together to prevent failures and, when a failure does occur, to reduce the consequences.

PHMSA regulates approximately 497,000 miles of onshore and offshore high pressure transmission pipelines in the United States. This number includes over 173,000 miles of hazardous liquid pipelines. As a first line of defense, PHMSA's comprehensive regulations seek to prevent the leading causes of pipeline failure such as corrosion, or to mitigate pipeline failures by addressing leak detection and leak containment using leading technology and best practices. Corrosion protection mechanisms include establishing safety standards for steel pipes that address protective coatings and design criteria to prevent corrosion and reduce latent defects. In addition, operators must implement corrosion prevention technologies, including specified cathodic protection systems, electric isolation of pipelines, and atmospheric corrosion prevention methods. With respect to pipeline leak protection, operators are required to deploy a system to detect and repair pipeline leaks as soon as possible. Finally, PHMSA also ensures that hazardous liquid pipeline operators are properly using, locating, and testing relief valves and emergency flow restricting devices or shut-off valves.[1]

As an added layer of defense, PHMSA mandates operators to establish Integrity Management Programs to further protect people and the environment in or near HCAs. These programs reflect a recognition that each pipeline is unique and has its own pipeline-specific risk profile dependent on its location, operating environment, the commodity being transported, and many other factors.

This testimony provides an overview of the Hazardous Liquid Pipeline Integrity Management Program requirements and PHMSA's oversight of these programs, highlighting how these programs intersect with regulations relating to corrosion, leak detection, and shut-off mechanisms.

I. Hazardous Liquid Pipeline Integrity Management Programs

Ten years ago, PHMSA mandated that hazardous liquid pipeline operators develop an Integrity Management Program to identify, assess, remediate, and validate, through comprehensive analyses, the integrity of hazardous liquid pipelines that could affect HCAs. This landmark set of regulations added a broad-reaching and fundamentally different approach to improving pipeline safety. Previous concepts of pipeline maintenance and inspection focused on the pipeline itself, investigating a pipeline's physical qualities, supporting systems, the administration of an operator's inspection program, and learning from accidents. The Integrity Management regulations supplement PHMSA's safety requirements with new requirements that are more intelligent, performance and process-oriented, setting expectations for operators and requiring them to identify and address risks unique to their pipelines. Hazardous Liquid Integrity Management Programs apply only to all hazardous liquid pipelines and carbon dioxide pipelines that could affect an HCA.

Fundamentally, an Integrity Management Program seeks to identify, prevent, and mitigate the potential consequences of failure of a specific pipeline at its location. The key components of an Integrity Management Program are: (1) HCA segment identification and initial risk analysis, (2) baseline and recurring pipeline assessment, (3) remediation of certain conditions and anomalies that could cause a pipeline failure, and (4) post-assessment risk analysis and establishment of specific plans to address known threats, including corrosion.

PHMSA requires each operator to know which parts of its pipeline system could affect an HCA in the event of a failure. High consequence areas are defined as (1) high population urbanized areas or other areas with a concentrated population, (2) unusually sensitive areas, and (3) commercially navigable waterways. PHMSA applied this definition to identify HCAs and made maps depicting those areas available to operators. Operators are responsible for independently evaluating information about the area around their pipeline to identify changes in circumstance that could result in new areas becoming HCAs.

The rule further requires an operator to consider how each pipeline segment could affect an HCA. In particular an operator must consider factors they identify as relevant to their operations and the HCA, as well as PHMSA-identified common factors, the Department's technical guidance, and reports by the National Transportation Safety Board (NTSB) and the Environmental Protection Agency. Next, operators must develop processes and tools to identify and analyze their pipeline's unique risks for failure. Among the mechanisms they employ to protect HCAs, operators must employ an effective means of detecting leaks on its pipeline system.

PHMSA's Integrity Management Program emphasizes prompt and remote detection of leaks through monitoring of operational parameters and engineered leak detection systems for all pipelines. Instead of requiring computer-based leak detection systems, PHMSA addresses existing leak detection system inadequacies with each operator by analyzing and evaluating each operator's leak detection capabilities for individual pipeline systems. PHMSA encourages, and in some cases requires, timely and comprehensive adoption and application of particular technology commensurate with the system-specific needs of each operator. The use of appropriate leak detection technologies enhances an operator's ability to detect and repair hazardous liquid pipeline leaks at the soonest possible time to mitigate any damages and risks. Earlier this year, PHMSA published an advisory bulletin informing all operators of PHMSA's expectations regarding pipeline leak detection systems, not just those operating in HCAs. The bulletin stated that the operating plans and procedures required by the pipeline safety regulations should include an engineering analysis to determine whether a computerized leak detection system is necessary to improve leak detection performance and line balance processes. The advisory resulted in the closure of an NTSB recommendation.

PHMSA's regulations set requirements for installing and locating valves during a pipeline's construction, as well as periodic testing. PHMSA requires operators to install relief or shut-off valves as appropriate to protect HCAs. In making the determination of where or whether to install shut-off or relief valves, an operator must consider the following factors:

(1) the swiftness of leak detection and pipeline shutdown capabilities; (2) the type of commodity being carried; (3) the rate of a potential leakage; (4) the volume that can be released; (5) the topography or pipeline profile; (6) the potential for ignition; (7) the proximity to power sources; (8) the location of nearest response personnel; (9) the specific terrain between the pipeline segment and the high consequence area; and (10) the benefits expected by reducing the spill size. While the regulations provide operators with discretion in locating shut-off valves, PHMSA inspectors ensure that the operators have adequately considered all appropriate considerations and placed shut-off valves appropriately. This determination has given operators as well as responders a greater awareness of the areas most susceptible to damage from pipeline failures. As a result, there is now a better understanding of where these most vulnerable areas are located and where additional protection is warranted.

After the operator has identified the HCAs and the potential risks, PHMSA requires operators to assess conditions on the relevant pipelines. Figure 1 shows the most common types of assessment methods used to evaluate the risks unique to each pipeline.

With respect to corrosion, the Integrity Management Program requires specific testing. An operator must use inline inspection tools and/or other approved inspection methods to identify corrosion and other deformities of the pipeline so that problem areas can be corrected and mitigating measures can be taken before a pipeline failure occurs. Figure 2 shows a downward trend of corrosion related accidents since the implementation of the Integrity Management Program and associated corrosion control regulations.

The Integrity Management Program imposed deadlines on hazardous liquid pipeline operators to complete an initial baseline assessment of pipelines that could affect HCAs. Approximately 44 percent of the hazardous liquid pipeline mileage falls within an area that could affect an HCA. Operators of large pipeline systems were required to complete their initial baseline assessments of all pipeline segments that could affect an HCA by March 31, 2008. Small operators with less than 500 miles were required to complete baseline assessments of their systems by February 17, 2009. PHMSA's inspections and the mandated annual reporting by operators (certified by company executives) have shown that these deadlines were met. Operators are now conducting the second round of assessments of these same pipeline segments. This assessment will increase our overall knowledge about the condition of the nation's pipelines in and beyond HCAs. Operators have internally inspected, pressure tested, or otherwise assessed approximately 86 percent of the total hazardous liquid pipeline mileage, well beyond areas designated as those that could affect an HCA.

Operators must remediate anomalies identified during their assessments that meet a certain defined criteria in a timely manner. PHMSA has defined the remediation timeframes for various types of anomalies. To date, Integrity Management Program pipeline assessments have revealed over 35,000 dangerous conditions within HCAs that pipeline operators have remediated. An additional 78,000 anomalies that were identified, but were not categorized as constituting dangerous conditions have been remediated, many of which were outside of HCAs. Those anomalies were not required to be repaired by the Integrity Management Program, but were discovered and proactively remediated as a result of assessments. Figure 4 depicts the number of repairs completed over and above those required by the Integrity Management Program regulations.

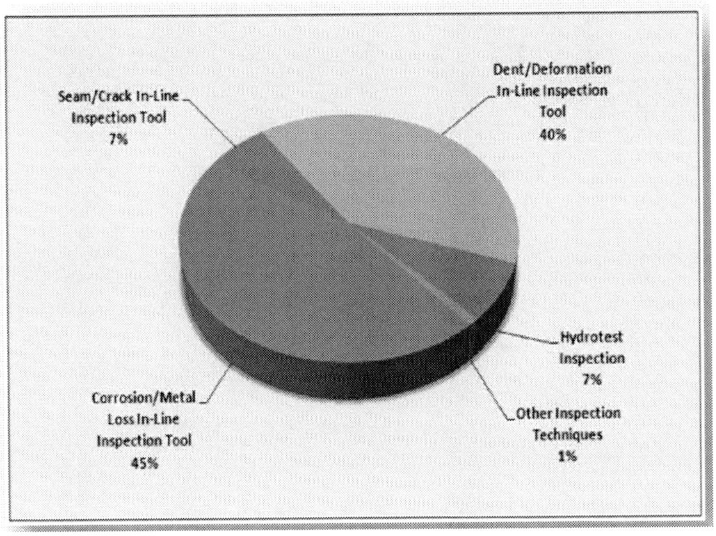

Figure 1. Types of Pipeline Inspections under the IM Rule 2001-2009.

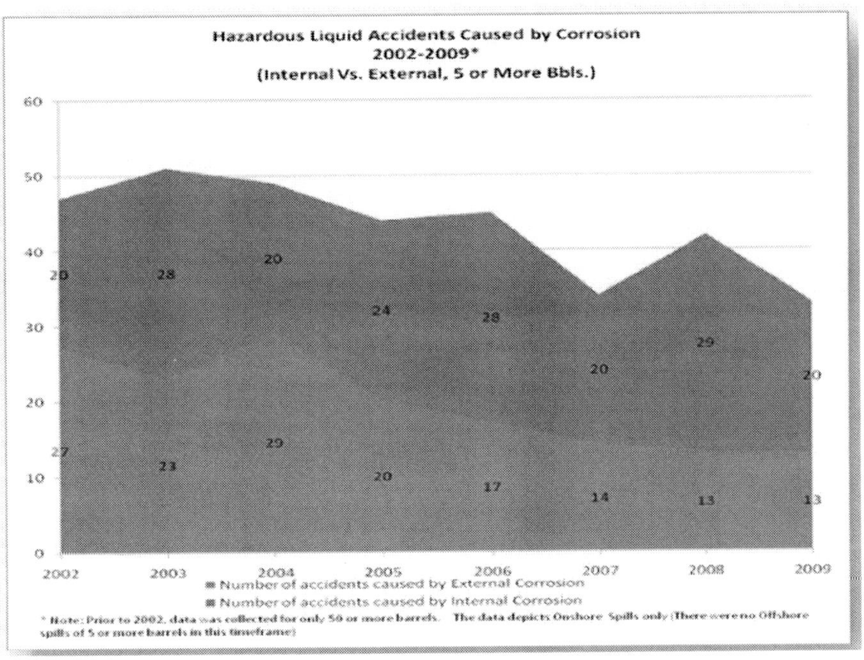

Figure 2. Onshore Hazardous Liquid Corrosion Accident Trends - 2002-2009.

After the initial assessment and remediation efforts, the Integrity Management Program requires operators to establish systematic approaches for using risk assessments to identify and implement additional preventive and mitigative measures. For example, operators must develop and implement a plan to mitigate the potential effects of corrosion if their assessments indicate corrosion risks. Pipeline operators are required to monitor the effectiveness of their preventive and mitigative programs, and re-evaluate the program if there are changed circumstances.

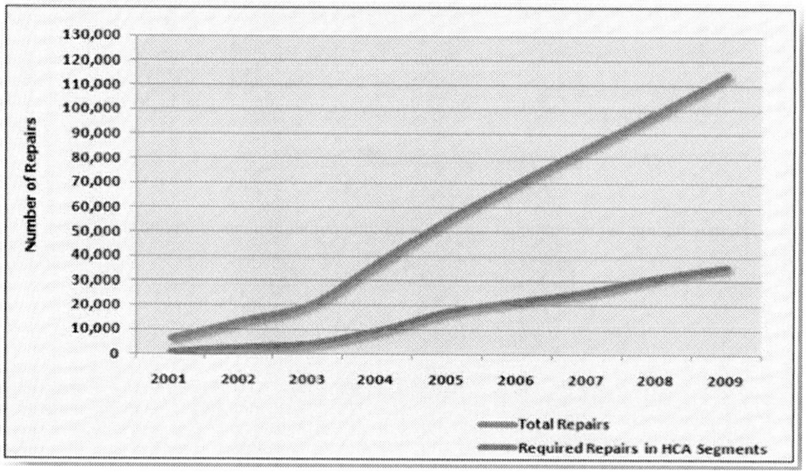

Figure 4. Repairs Made Under the Liquid IM Rule 2001 -2009 (Cumulative).

The Integrity Management Program goes beyond simply assessing pipeline segments and repairing defects. Improving operator Integrity Management Programs, analytical processes, and their application is also a critical objective. The ability to integrate and analyze threat and integrity related data from many sources is essential for proactive safety management.

II. PHMSA'S OVERSIGHT AND ENFORCEMENT OF INTEGRITY MANAGEMENT PROGRAMS

PHMSA developed an inspection program to assure compliance with the Integrity Management Program requirements. It also developed a comprehensive set of inspection protocols that not only check for compliance with the regulation's prescriptive requirements, but also support a detailed audit of an operator's management and analytical systems, processes, and practices on pipeline integrity. PHMSA developed a specialized training program for federal and state inspectors to inspect these protocols. To date, 383 Federal and State inspectors have been trained to inspect Integrity Management Programs. The Integrity Management Program of every operator PHMSA regulates has been inspected at least once. All large hazardous liquid pipeline operators have been inspected a second time to be sure they are continuing to manage pipeline integrity and making progress in building robust and effective Integrity Management Programs.

When operators fall short of meeting the requirements for Integrity Management Program development, PHMSA takes enforcement action to accelerate program development and address program deficiencies with the individual operators. PHMSA issued enforcement letters in 85% of all its Integrity Management inspections. PHMSA also has not hesitated to exercise its civil penalty authority when violations of the rule's requirements occur.

PHMSA also ensures that operators comply with corrosion standards though inspection and aggressive enforcement. PHMSA has initiated 272 enforcement cases resulting from 657 probable violations or procedural inadequacies involving corrosion problems since 2000. It has proposed $1,798,950 in fines on those actions. In addition, many of the Integrity Management enforcement actions addressed corrosion issues.

Similarly, PHMSA inspectors have identified a number of issues related to an operator evaluation of their leak detection capabilities using the Integrity Management inspection protocols. PHMSA has initiated enforcement actions, or formally documented its concerns, with respect to approximately 40 percent of hazardous liquid pipeline operators to date. In response to the enforcement actions, operators are required to submit revised procedures to correct inadequacies in their leak detection evaluations. Operators must then evaluate (or reevaluate) their leak detection capabilities in accordance with these corrected procedures. Before a case is closed, PHMSA reviews the revised procedures, and determines that the revisions satisfactorily address identified issues.

Transparency has been a hallmark of PHMSA's regulatory oversight. PHMSA has a website for Implementing Integrity Management to provide information to the public on the rule as well as PHMSA's oversight of the program. This publicly accessible website includes hundreds of Frequently Asked Questions to explain the rule's provisions and PHMSA's expectations. This resource also provides the inspection protocols, an Integrity Management fact sheet, a glossary, a flow chart of the Integrity Management process, reference documents, and industry performance measures. The public also has access to information on enforcement cases stemming from PHMSA's Integrity Management inspections via the website. PHMSA has taken unprecedented steps to inform the public and all other stakeholders about the protections provided by the Integrity Management Program and PHMSA's oversight.

PHMSA looks forward to seeing Integrity Management Programs continue to mature and yield results. With this in mind PHMSA will continue to look at performance measures and ways we can improve the data that we collect. Having better data will enable PHMSA to make risk based informed regulatory decisions.

With anticipated increases in transportation of new products like ethanol, hydrogen, carbon dioxide, and potentially other bio-fuels, PHMSA is working to ensure a solid regulatory framework to prevent accidents and ensure safety. PHMSA is committed to taking whatever steps are necessary to ensure that such transportation will be conducted safely. We coordinate with other federal agencies to forecast the transportation implications from the inception of marketing new fuels, as part of a systemic oversight process. We coordinate with other countries to benefit from their experience. We continue to work with individual operators, identifying safety concerns that must be satisfied, both with the infrastructure and with the surrounding community. We continue to collaborate with the pipeline industry, emergency response organizations, and others to investigate and solve technical challenges.

In closing PHMSA looks forward to working with Congress to address issues related to hazardous liquid pipeline safety, including finding ways to prevent pipeline failures and mitigate the effect of any failure. PHMSA very much appreciates the opportunity to report on the status of our progress in preventing spills from hazardous liquid pipelines.

Thank you. I would be pleased to answer any questions you may have.

End Notes

[1] *Relief valves are a type of valve used to limit the pressure in a pipeline system. Shut-off valves are automatic or remote controlled valves that close the flow on pipelines.*

In: Pipeline Safety and Security
Editor: Jonathan T. Miller and Lisa M. Gurton

ISBN: 978-1-61209-147-1

Chapter 5

"The Safety of Hazardous Liquid Pipelines: Regulated vs. Unregulated Pipelines" Hearing, before the Subcommittee on Railroads, Pipelines, and Hazardous Materials

Jonathan T. Miller and Lisa M. Gurton

Purpose of the Hearing

The Subcommittee on Railroads, Pipelines, and Hazardous Materials is scheduled to meet on Tuesday, June 29, 2010, at 2:00 p.m., in room 2167 of the Rayburn House Office Building to receive testimony on the safety of hazardous liquid pipelines. The hearing will focus on what pipelines are regulated; what pipelines are exempted from regulation; and any gaps that may exist in the current statute or regulations.

Background

According to the Pipeline and Hazardous Material Safety Administration (PHMSA), there are more than 170,000 miles of onshore and offshore hazardous liquid pipeline (about 200 operators) in the United States, which carry more than 75 percent of the nation's crude oil and around 66 percent of its refined petroleum products. Economists predict hazardous liquid pipeline mileage will grow over the next 25 years.[1] Of the more than 170,000 miles of hazardous liquid pipeline, about 55,000 miles are major crude oil trunk lines, which range in diameter from about eight inches up to 48 inches. Associated with these trunk lines in several locations is significant crude oil tankage, and about 30,000 to 40,000 miles of crude gathering lines which are smaller lines that gather the oil, gas, and water from many wells, both onshore and offshore, and connect to the larger trunk lines. In addition, there are about 95,000 petroleum product lines, flow lines/piping associated with well operations, and produced water pipelines (containing contaminated water following oil, gas, and water separation).

I. Who Has Jurisdiction Over Pipeline Safety

Within the Department of Transportation (DOT), PHMSA's Office of Pipeline Safety is responsible for overseeing the safety of the Nation's pipeline system; from design specifications and construction procedures to operation, maintenance, and onshore spill response planning. PHMSA only has jurisdiction over transportation-related facilities; it does not have jurisdiction over drilling or production facilities.

PHMSA has about 200 full-time staff, including inspectors, to oversee the pipeline safety program. In addition to its own staff, PHMSA authorizes States to conduct oversight of intrastate and interstate pipelines in lieu of Federal oversight.[2] The States must certify annually to the Secretary that they have adopted, or are in the process of adopting, the Federal standards; are enforcing the standards; and are encouraging and promoting the establishment of damage prevention programs. Each annual certification must include a report that contains all accidents or incidents reported over the last year to the State involving a fatality, personal injury requiring hospitalization; property damage or loss of more than $50,000; any other accident the State considers significant; and a summary of the investigation by the State of the cause and circumstances surrounding the accident or incident. Each State must also submit the record maintenance, reporting, and inspection practices they conducted to enforce compliance with Federal safety standards, including the number of inspections of pipeline facilities the State made during the previous 12 months. Today, 17 States are certified to inspect intrastate hazardous liquid pipelines. [3] The Secretary is authorized to then make an agreement with those certified States to participate in the oversight of interstate pipeline transportation. Six States are authorized to conduct inspections for interstate hazardous liquid pipelines.[4]

II. Accidents and Incidents

Taken as a whole, releases from hazardous liquid pipelines cause few annual fatalities compared to other product transportation modes. According to PHMSA's website there were 100 significant hazardous liquid pipeline incidents (98 onshore and two offshore) in 2009, resulting in four fatalities, four injuries, and about $59 million in damages. Corrosion is the leading cause of all hazardous liquid pipeline incidents. However, PHMSA does not provide to the public information on all reported pipeline incidents, only serious and significant incidents. PHMSA defines "significant incidents" as an incident resulting in: (1) a fatality or injury requiring in-patient hospitalization; (2) $50,000 or more in total costs, measured in 1984 dollars; (3) a release of five barrels (210 gallons) or more of highly volatile liquid or 50 barrels (2,100 gallons) or more of other hazardous liquid; and (4) an unintentional fire or explosion. Such reporting by PHMSA may be misleading, as the statistics that PHMSA publishes may not include incidents that may be valuable in evaluating the safety of the industry. Upon request of Committee on Transportation and Infrastructure Majority staff,

PHMSA provided information on all reported pipeline incidents over the last five years. In 2009, there were 331 reportable hazardous liquid pipeline incidents; of those, 169 were spills involving five to 210 gallons of hazardous liquid, and 162 involved the release of 210 gallons or more. Only 100 of these 331 incidents are reported to the public on PHMSA's website.

Year	Total Incidents Reported to PHMSA	Total Incidents Reported to the Public on PHMSA's Website	Percentage of Hazardous Liquid Pipeline Incidents Not Reported to the Public
2005	363	129	64
2006	351	106	69
2007	328	107	67
2008	373	115	69
2009	331	100	69
Total	**1,746**	**557**	**68**

Source: PHMSA, Office of Pipeline Safety

Even with these figures, it is important to mention that not all pipeline incidents are reported to PHMSA. Some pipelines are exempt from reporting requirements, such as unregulated gathering lines and produced water lines. In addition, certain incidents are exempt from accident reporting requirements. PHMSA regulations only require reporting when an incident involves a fatality or injury; a release of more than five gallons of hazardous liquid or carbon dioxide; or $50,000 in property damage. Incidents involving a pipeline maintenance activity which result in a release of less than 210 gallons of hazardous liquid and are cleaned up promptly and confined to company property or pipeline right-of-way are also exempt from reporting.

Although pipeline releases have caused relatively few fatalities in absolute numbers, a single pipeline incident can be catastrophic. For example, a 1999 gasoline pipeline explosion in Bellingham, Washington, killed two children and an 18-year-old man, and caused $45 million in damage to the city water treatment plant and other property. In 2006, corroded pipelines on the North Slope of Alaska leaked more than 200,000 gallons of crude oil in an environmentally sensitive area. On May 25, 2010, during a scheduled shut down of the 800-mile Trans-Alaska Pipeline for maintenance, relief Tank 190 at Pump Station #9 was overfilled and crude oil spilled into secondary containment. PHMSA estimates that the amount of oil spilled is about 5,000 barrels (210,000 gallons).

These consequences can be significantly reduced with adequate Federal pipeline safety requirements and enforcement.

III. How Does Phmsa Regulate?

In addition to the rules set forth in Executive Order 12866 and Office of Management and Budget regulations, in 1996, the "Accountable Pipeline Safety and Partnership Act of 1996" (P.L. 104-304) set forth standards that the Secretary must consider when prescribing a regulation relating to pipeline safety, and a procedure that required the Secretary to submit a risk assessment to two technical advisory committees prior to initiating a rulemaking proceeding.[5] The Technical Hazardous Liquid Pipeline Safety Standards Committee (Committee) is composed of 15 members appointed by the Secretary with "the technical aspect of transporting hazardous liquid or operating a hazardous liquid pipeline facility." Each member must have experience in transporting hazardous liquid or operating hazardous liquid pipeline facilities, or be technically qualified in at least one field of engineering applicable to transporting hazardous liquid or operating a hazardous liquid pipeline facility.

The risk assessment submitted by the Secretary to the Committee must identify all regulatory and non-regulatory options that the Secretary considered; the costs and benefits of the proposal; an explanation of the reasons for the selection of the proposed standard in lieu of alternative proposals; and include identification of any technical data the risk assessment was based on. The Committee then reviews the assessment and submits a report to the Secretary that includes an evaluation of the merit of the proposal; and any recommended options relating to the risk assessment information or the proposed standard. The Secretary is then required to review the report and provide a written response to the Committee concerning the comments and recommended alternatives.

In addition, the law prohibits the Secretary from proposing or issuing any standard relating to pipeline safety unless the benefits outweigh the costs. The law states: "The Secretary shall propose or issue a standard under this Chapter only upon a reasoned determination that the benefits of the intended standard justify its costs." The only exception to this, according to PHMSA, is a mandate from Congress, a proposal from the technical advisory committees, or as a result of negotiated rulemaking.

Following enactment of the 1996 law, PHMSA established a working group to develop a collaborative process for performing cost-benefit analyses, which consisted of a few government representatives and representatives from the regulated industry – the American Petroleum Institute, the Gas Research Institute, the American Gas Association, the Interstate Natural Gas Association of America, the American Public Gas Association, and a number of hazardous liquid, natural gas distribution, and natural gas transmission companies. The working group developed a documented framework to work collaboratively on future rulemakings, which to this day is what PHMSA uses to develop its proposals.

IV. Exemptions from Pipeline Safety Regulation

On November 29, 2009, a BP p.l.c. (BP) oil field operator doing a routine inspection at a drill site in the Prudhoe Bay (Alaska) oil field found a 24-inch jagged rupture in a pipeline which began pouring oil and water on the snow-covered tundra. The on-scene coordinator for the state Department of Environmental Conservation reported that "the breach on the bottom of the pipe was the biggest he had ever seen and indicative of the incredible pressure the pipeline was under when it split." Officials found that massive ice plugs had formed inside the pipe which caused BP to stop operating it. Pressure then built up until the pipeline ruptured, spilling 46,000 gallons of crude and produced water (contaminated water following oil, gas, and water separation). The pipeline was just six inches in diameter and 18 inches long, and it is not regulated by the Federal Government. It is regulated by the State, but the State has failed to enforce its own regulations.

A number of pipelines are exempt from the PHMSA's safety regulations, including: (1) certain onshore gathering lines and gathering lines located in an inlet of the Gulf of Mexico; (2) certain rural low-stress pipelines; (3) pipelines subject to safety regulations of the US Coast Guard; (4) certain low-stress (i.e. low-pressure) pipelines that serve refining, manufacturing, or truck, rail, or vessel terminal facilities; (5) certain offshore pipelines that transport hazardous liquid or carbon dioxide in State waters; (6) certain pipelines on the Outer Continental Shelf that transport hazardous liquid or carbon dioxide; (7) transportation of a hazardous liquid or carbon dioxide through onshore production (including flow lines), refining, or manufacturing facilities or storage or in-plant piping systems associated with such facilities; (8) certain pipelines transporting carbon dioxide; and (9) certain produced water lines (post oil/gas/water separation lines carrying briny water with oil and gas contaminants).

Numerous safety and environmental groups have requested that PHMSA review these exemptions, stating as recently as 2006: "[These] exemptions result in regulatory coverage that is piecemeal at best and confusing, difficult to implement and enforce, and inadequate at worst. These exemptions were developed or modified over many years, and often are where pipeline release problems lie today. Additionally, the exemptions weaken the public's confidence in PHMSA's ability to ensure pipeline safety."[6]

V. Low-Stress Pipeline Exemption

Perhaps the best example of the importance of regulating pipelines came in March 2006, when an internal corrosion on a 34-inch low-stress pipeline, owned by BP Exploration, which at the time was unregulated by PHMSA because it was a low-stress pipeline, caused a 5,000 barrel crude oil spill (212,252 gallons spilled) on the North Slope of Alaska. The oil spill was the worst in the history of oil development on Alaska's North Slope, and went undetected for five days before a BP oilfield worker detected the scent of hydrocarbons during a drive through the area. It was later learned by Federal investigators that BP had ignored at least four alarms on its Supervisory Control and Data Acquisition (SCADA) system – a computer system used for monitoring and controlling the pipeline – indicating there was a leak.

A few months later, in August 2006, a second leak was discovered while BP was inspecting the Eastern Operating Area segment of the pipeline. Field inspection of the leak site revealed multiple holes at a single location, contributing to an estimated spill of approximately 1,000 gallons of processed crude oil.

The cause of the leaks was internal corrosion. Federal investigators found that BP had not established a regular maintenance pigging (cleaning pig) or internal inspection (smart pigging) program on the pipelines. In fact, BP had never run cleaning pigs on the Eastern Operating Area pipelines since they took over operation of the pipelines in 2000. BP's predecessor, ARCO Alaska, had last cleaned and smart pigged the lines in 1992 and then suspended smart pigging of the Eastern Operating Area pipeline when residues, waxes, and calcium carbonate deposits clogged the Trans Alaska Pipeline strainers. Before the 2006 spill, an internal inspection of the Western Operating Area pipeline, which BP has always operated, was last performed in 1998 using a high-resolution magnetic flux leakage tool. According to PHMSA at the time, these should have been indications to BP that the lines needed significant cleaning and were at risk of rupturing. Once BP was forced to clean the lines after the Alaska spills, the lines were so corroded that the pigs actually got stuck during cleaning operations. In the end, PHMSA ordered BP to completely replace the lines. Replacement was completed in December 2009.

As a result of the spills, Congress directed PHMSA in the Pipeline Inspection, Protection, Enforcement and Safety Act of 2006 (Public Law 110-432) to issue a rulemaking that subjects all low-stress (i.e. low-pressure) hazardous liquid pipelines to the same standards and regulations as all other hazardous liquid pipelines. Unfortunately, it wasn't the first time PHMSA had heard the need to eliminate the low-stress pipeline exemption (which dates back to 1969).

- In 1988, the National Association of Pipeline Safety Representatives sent DOT a resolution asking that the low-pressure exemption be eliminated.
- In 1990, DOT asked for comments on "whether and to what extent" to remove the exemption from its regulations.
- In 1993, Volpe National Transportation Systems Center reported to DOT that there were 20,000 miles of onshore rural gathering lines and 22,000 miles of unregulated low-pressure transmission pipelines. Volpe estimated that 38 percent of the 22,000 miles were near a populated area or a navigable waterway, leaving 15,000 miles of low-pressure transmission pipelines unregulated.
- As a result, in 1994, DOT adopted a final rule to regulate a portion of low-stress pipelines: those in non-rural areas and areas currently used for commercial navigation. All other low-stress pipelines remained unregulated, at least until Congress mandated that they be regulated following the 2006 BP spills in Alaska.

On June 3, 2008, in response to the congressional mandate, PHMSA published a Final Rule regulating only 803 miles of large diameter, low-stress pipelines. PHMSA stated that in the Final Rule that it needed more time to gather data about the universe of unregulated low-stress pipelines, and that it would come back in a second rulemaking and regulate all other applicable low-stress pipelines.

On June 23, 2010, PHMSA issued a Notice of Proposed Rulemaking (NPRM) to regulate all the other low-stress pipelines. Unfortunately, the Final Rule issued in 2008 and the NPRM does not comply with the congressional mandate. Congress directed PHMSA to subject all low-stress pipelines to the same standards and regulations as all other hazardous liquid pipelines. Under existing regulations, hazardous liquid pipelines must conduct a comprehensive analysis of the integrity of all of their pipelines that "could affect" a high-consequence area (HCA). HCAs are populated areas, commercially navigable waterways, and unusually sensitive areas (areas requiring extra protection because of the presence of sole source drinking water, endangered species, or other ecological resources that could be damaged by oil leaks). Problems found in the pipeline as a result of these comprehensive analyses must be repaired within certain specified timeframes.

Instead of subjecting low-stress pipelines to the same "could affect a high-consequence area" standard as all other hazardous liquid pipelines, PHMSA adopted a specific mileage buffer. PHMSA states that low-stress pipelines, unlike the more stringent requirements of other hazardous liquid pipelines, within a half-mile of an unusually sensitive area would have to conduct the comprehensive integrity management assessments. Low-stress pipelines outside the half-mile buffer area would be exempt from those requirements. PHMSA maintains that this is because there has never been an incident involving a low-stress pipeline outside of a half-mile buffer area. Safety advocates argue that such a buffer fails to address the potential for spilled hazardous liquids to move to environmentally-sensitive areas through water or watersheds from farther than a half-mile away. Nevertheless, PHMSA said that using a "could affect" determination would be "burdensome" for industry.

In addition, PHMSA stated that it would consider exempting certain pipelines (through the special permit process) from the new rule due to economic hardship: those operators who transport crude oil from a production facility and operate at a flow rate less than or equal to 14,000 barrels per day as long as the operators maintain that they may abandon or shut-down the pipeline as a result of the economic burden of complying with integrity management assessment requirements.

VI. Exemptions Through Issuance of Special Permits

Under current law, PHMSA may waive or modify compliance with an existing regulation if the pipeline operator requesting such a waiver, known as a special permit, can demonstrate the need for the waiver and PHMSA determines that granting the waiver is consistent with pipeline safety.[7] The operator is required to provide PHMSA with (1) a detailed description of the pipeline facilities for which the special permit is sought; (2) a list of the specific regulation(s) from which the operator seeks relief; (3) an explanation of the unique circumstances that the operator believes make the applicability of that regulation or standard (or portion thereof) unnecessary or inappropriate for its facility; (4) a description of any measures or activities the operator proposes to undertake as an alternative to compliance with the relevant regulation, including an explanation of how such measures will mitigate any safety or environmental risks; (5) a description of any positive or negative impacts on affected stakeholders and a statement indicating how operating the pipeline pursuant to a special permit would be in the public interest; (6) a certification that operation of the operator's pipeline under the requested special permit would not be inconsistent with pipeline safety; and (7) any other information PHMSA may need to process the application including environmental analysis where necessary.

If PHMSA determines that the application complies with the requirements for a special permit and that the waiver of the relevant regulation or standard is not inconsistent with pipeline safety, PHMSA may grant the application, in whole or in part. Conditions may also be imposed on the granting of the waiver if PHMSA concludes they are necessary to assure safety, environmental protection, or are otherwise in the public interest.

An example of a special permit request is Keystone XL tar sands pipeline, which would bring oil from Canada to the Gulf of Mexico. TransCanada has applied to PHMSA for a special permit to design, construct, and operate a more than 1,300-mile, 36-inch pipeline using a design factor (thinner pipe) and higher maximum operating pressure than is permitted under existing regulations.

VII. Integrity Management for Hazardous Liquid Pipelines

On February 1, 2000, in the wake of several pipeline ruptures in Bellingham, Washington; Simpsonville, South Carolina; Reston, Virginia; and Edison, New Jersey, PHMSA issued a Final Rule requiring pipeline operators to develop and implement an integrity management program, which requires operators to first determine what pipelines are located in a HCA. According to PHMSA, HCAs represent 44 percent of the total hazardous liquid pipeline mileage.

Once the HCAs were identified, operators were required to comprehensively inspect the integrity of their pipelines that, in the event of a failure, could affect an HCA. Based on those evaluations, the operators were required to take prompt action to repair any defects that could reduce a pipeline's integrity.

All baseline assessments for operators with more than 500 miles of pipeline were to be completed by March 31, 2008; all others were to be completed by February 15, 2009. According to PHMSA, the assessments revealed thousands of hazardous liquid pipeline defects: 35,000 were reported. More than 3,800 serious hazardous liquid pipeline defects had to be repaired immediately; another 14,000 hazardous liquid defects had to be repaired within a 60- to 180-day time period.[8] Information on the nature, or extent, of the defects is not publicly available. Only the number of defects identified and number of repairs made are reported to PHMSA and then reported to the public.

Safety advocates maintain – particularly given the large number of defects identified as a result of the assessments – that PHMSA should expand the scope of the integrity assessments to require pipeline operators to evaluate the integrity of their pipelines outside of HCAs. Pipeline operators, however, maintain that while they are only required to assess pipelines that could affect HCA's, in practice they evaluate a much greater percentage of pipelines when they conduct these assessments. According to the operators, this is due largely to the practical constraints associated with running in-line inspection tools, such as smart pigs; because of the location of the launchers and receivers used to insert and remove smart pigs from the pipeline, relatively long sections of pipeline are inspected when these tools are used. These sections generally contain portions of the pipeline that can affect HCAs and portions of the pipeline that do not affect HCAs. Thus, while conducting assessments of the portions of their lines that affect HCAs, operators running smart pigs also obtain data on the condition of their pipelines in other areas and take action to assure the integrity of those sections outside of HCAs.

Safety advocates, however, maintain that although larger areas may be assessed, Federal regulations mandating reporting any defects identified or repairs made are applicable only to those pipeline segments that could affect HCAs. Any reporting to PHMSA beyond HCAs is strictly voluntary.

VIII. Are There Gaps in Existing Regulations?

Witnesses are expected to also discuss any perceived gaps in existing safety regulations, including safety recommendations of the National Transportation Safety Board that have not yet been adopted.

End Notes

[1] The outlook during the first 25 years of the 21st Century is for U.S. petroleum product demand to increase 9.5 million barrels per day (48 percent) with two-thirds of the growth being for transportation fuels. During that time, inland crude production is expected to decline 900 thousand barrels per day, mostly in Texas, Louisiana, Oklahoma, and the Rocky Mountain States, while Gulf of Mexico production likely will increase by 500 thousand barrels per day. The forecast shows refining capacity growing 3.3 million barrels per day, mostly in Texas and Louisiana. The outlook would necessitate imports growing substantially, with crude up four million barrels per day and refined products up 6.3 million barrels per day. During the same period, significant growth is expected in the petrochemical industry. Richard A. Rabinow, *The Liquid Pipeline Industry in the United States: Where It's Been, Where It's Going*, prepared for the Association of Oil Pipe Lines (April 2004).

[2] 49 U.S.C. § 60105.

[3] Alabama, Arizona, California (Fire Marshal), Indiana, Kentucky, Louisiana, Maryland, Minnesota, Mississippi, New York, New Mexico, Oklahoma, Pennsylvania, Texas, Virginia, Washington, West Virginia (PHMSA, *CY 2010 State Program Certification/Agreement Status* (December 2009)).

[4] Arizona, California, Minnesota, New York, Virginia, and Washington (PHMSA, *CY 2010 State Program Certification/Agreement Status* (December 2009)).

[5] 49 U.S.C. § 60102.

[6] Pipeline Safety Trust, comments filed with PHMSA to the docket on the low-stress pipeline rulemaking.

[7] 49 U.S.C. § 60118.

[8] PHMSA, PowerPoint presentation, *The Pipeline Inspection Program*, prepared upon request of House Transportation and Infrastructure Committee Majority Staff (March 2010).

In: Pipeline Safety and Security
Editor: Jonathan T. Miller and Lisa M. Gurton

ISBN: 978-1-61209-147-1

Chapter 6

Written Statement of Cynthia L. Quarterman, Administrator, Pipeline and Hazardous Materials Safety Administration, U.S. Department of Transportation, before the Subcommittee on Railroads, Pipelines, and Hazardous Materials, Hearing on "Hazardous Liquid Pipeline Safety: Regulated vs. Unregulated"

Cynthia L. Quarterman

Chairwoman Brown, Ranking Member Shuster, members of the Subcommittee, thank you for the opportunity to appear today to discuss the Pipeline and Hazardous Materials Safety Administration's (PHMSA) oversight of America's hazardous liquid pipeline system. The top priority for Secretary LaHood and all of us at PHMSA is safety. We have strong commitments to reducing transportation risks to the public and environment. Our Nation's reliance on the safe and environmentally sound transportation of energy fuels and hazardous materials is increasing. PHMSA's safety oversight of the pipeline network that delivers these products is providing critical protections for the American people. The diagram below illustrates how hazardous liquids move through our pipelines from the well to the consumer:

For years, PHMSA has worked aggressively to utilize the authority given to it by Congress to enhance the safety of hazardous liquid pipelines. To do this, the agency has used responsible and methodical approaches to focus on high-risk infrastructure issues first, and provide effective solutions through enforcement and rulemakings. This tiered approach has helped PHMSA devise and implement effective rulemakings, like the ones now in place for Low Stress lines. With Congressional support, PHMSA has also built and deployed a consensus-based, collaborative research and development program that is bringing new technologies to market and helping to strengthen and maintain pipeline integrity.

Hazardous Liquid Pipelines – From Well to Consumer

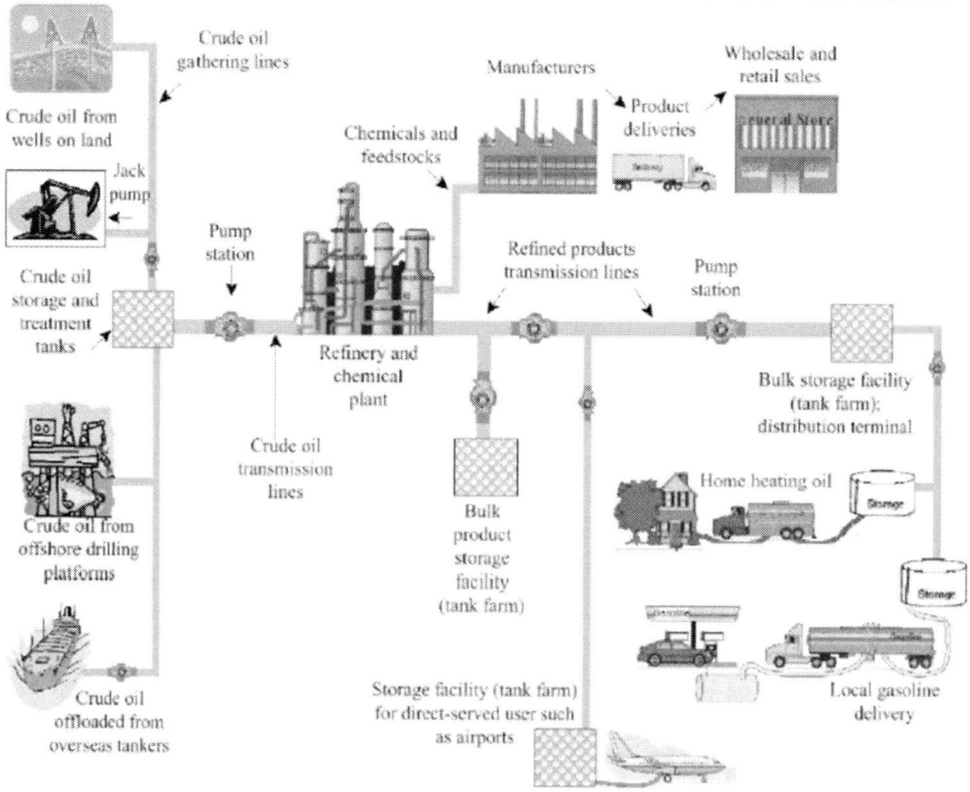

These initiatives have been successful. PHMSA has driven down the number of serious pipeline incidents over the past 20 years, while all the traditional measures of risk exposure have been rising – population, energy consumption, and pipeline ton-miles. Pipeline incidents involving death or injury have declined by 50% over the last two decades. As indicated in the chart below, we aim to continue this long-term trend.

The Pipeline Safety Law and regulations apply to the transportation of hazardous liquids by pipeline under the statutory jurisdiction of PHMSA and the States. That combined jurisdiction is broad and covers any pipeline in or affecting interstate or foreign commerce, including wholly intrastate pipelines and pipelines located on the Outer Continental Shelf (OCS). PHMSA jurisdiction encompasses any pipeline that transports a highly volatile liquid (e.g., propylene, ethylene, butylene, and anhydrous ammonia) or other hazardous liquids through a non-gathering line pipeline that has a maximum operating pressure greater than 20 percent of its specified minimum yield strength. In addition, PHMSA has jurisdiction over any pipeline that crosses a waterway used for commercial navigation, certain onshore petroleum gathering lines, and certain hazardous liquids or carbon dioxide low-stress pipelines.

PHMSA can only provide oversight and ensure the safety of hazardous liquid pipelines under its jurisdiction. As requested by this Subcommittee, I will provide a brief description of the statutory and regulatory authorities held and exercised by PHMSA and its State partners to oversee the safety of hazardous liquid pipelines. My testimony today will explain

PHMSA's authority over hazardous liquid pipelines and how it uses thisauthority to minimize safety risks.

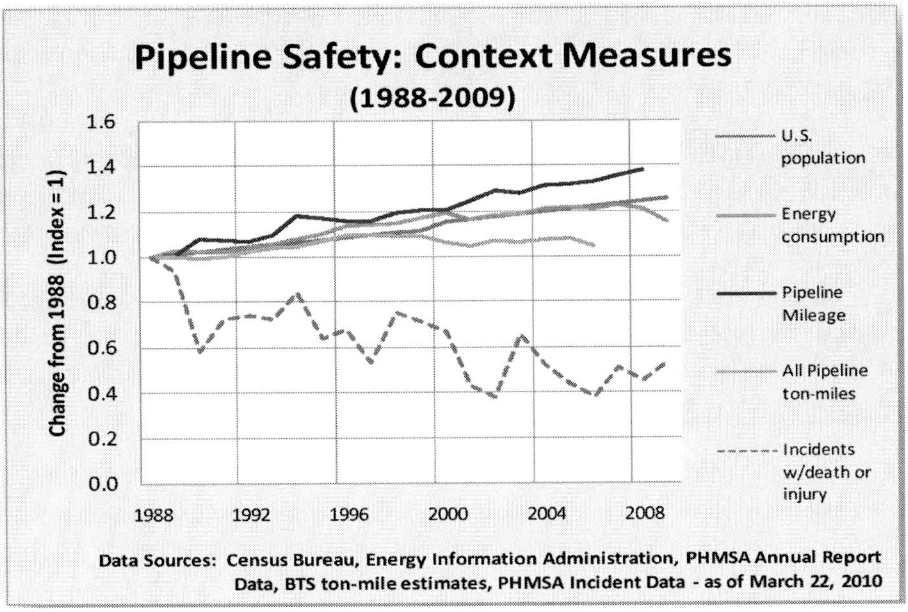

I. CONSTITUTIONAL AUTHORITY

The Commerce Clause of the U.S. Constitution is the authority underlying the Pipeline Safety Laws.[1] It permits federal regulation of the transportation of hazardous liquids bypipeline . Pursuant to that authority, Congress may mandate federal regulation of the use of the channels of interstate commerce, the instrumentalities of and persons or things in interstate commerce, and any activity that has a substantial effect on interstate commerce[2].

II. STATUTORY AUTHORITY AND PHMSA'S REGULATORY EXCLUSIONS

A. Congress Provided PHMSA and States Authority to Regulate the Transportation of Hazardous Liquids by Pipeline

1. PHMSA Has Broad, but not Unlimited, Statutory Authority to Regulate the Transportation of Hazardous Liquids by Pipeline.

PHMSA has statutory authority over "transporting hazardous liquids." "Hazardous liquid," is defined in the Pipeline Safety Laws as petroleum or a petroleum product, or any substance in a liquid state that the Secretary of Transportation decides may pose an unreasonable risk to life or property, including carbon dioxide.[3] "Transporting hazardous liquid," is defined in the Pipeline Safety Laws as the movement (or storage incidental to such movement) of a hazardous liquid by pipeline in or affecting interstate commerce; but excluding the movement of a hazardous liquid through gathering lines in rural areas; onshore production, refining, or manufacturing facilities; or storage or in-plant piping facilities associated with onshore production, refining or manufacturing facilities.[4]

Congress has further defined PHMSA's jurisdiction by including a statutory meaning of gathering lines. A "regulated gathering line" must be defined in regulation by PHMSA based upon consideration of certain physical and functional factors. These factors include location, length from the well site, operating pressure, throughput, and composition of the product transported. Whereas a crude oil gathering line that is less than 6 inches in diameter, operates at low pressure, and is located in a rural area that is not unusually sensitive to environmental damage, is explicitly excluded from regulation by statute.[5]

Moreover, low-stress hazardous liquid pipelines (i.e., those operating at a relatively low pressure) that are regulatedby the U.S. Coast Guard, or that serve refining, manufacturing, or truck, rail, or vessel terminals, which are less than one mile long and do not cross an offshore or commercially navigable waterway, are excluded from PHMSA oversight (at least until PHMSA completes the Low Stress Rulemaking).[6] The transportation of hazardous liquids or carbon dioxide through onshore production, refining, or manufacturing facilities (and any associated storage or in-plant piping systems) is also excluded.[7] The Occupational Safety and Health Administration regulates some of these facilities to ensure safety of workers, and others are regulated by State agencies. These facilities and associated piping are considered non-transportation-related pursuant to Executive Order 12777 and are regulated by the Environmental Protection Agency(EPA). [8]

While PHMSA is responsible for ensuring the safety of hazardous liquid pipeline transportation, it does not have the authority to determine the site or route of those facilities.[9] Other Federal agencies, including the Department of the Interior, the Department of State, the EPA, and State agencies make those decisions.

PHMSA's rulemaking authority is also prescribed by the Pipeline Safety Laws. Specifically, PHMSA must consider certain factors in those proceedings by statute, including any relevant hazardous liquid pipeline safety information, the appropriateness and reasonableness of any proposed standard, and the reasonably-identified costs and benefits of any new regulation,[10] and except where otherwise provided by statute, can only issue a regulation if its benefits justify its costs.[11]

Finally, there are other laws of more general applicabilitythat PHMSA must comply with in exercising its regulatory responsibilities, for example: (1) the National Environmental Policy Act of 1969,[12] a statute that requires Federal agencies to consider the environmental impacts of and proposed alternatives to certain regulations, (2) the Paperwork Reduction Act of 1995,[13] a statute that imposes certain requirements on the collection of information, and (3) the National Technology Transfer and Advancement Act of 1995,[14] a statute that encourages federal agencies to use consensus industry standards.

2. The States Have Statutory Authority to Regulate the Transportation of Hazardous Liquids by Pipeline.

Congress has preserved a role for the States in regulating the intrastate transportation of hazardous liquids by pipeline. In particular, a State is allowed to regulate exclusively a pipeline if located wholly within its borders, provided that State has a current certification or agreement with PHMSA and has adopted standards that are compatible with the minimum federal requirements.[15] A State may also serve as PHMSA's agent for purposes of inspecting interstate hazardous liquid pipeline facilities.[16] States cannot adopt or apply any of its own regulations to those facilities.[17]

B. There are Hazardous Liquids Pipelines that PHMSA Does Not Have Statutory Authority to Regulate

As is obvious, the Pipeline Safety Regulations do not apply to the transportation of hazardous liquids where precluded by statute. That includes the movement of hazardous liquids by pipeline through certain gathering lines in rural areas, and storage associated with onshore production, refining or other manufacturing facilities or non-pipeline modes of transportation.[18] These pipelines typically move unprocessed crude oil from producing well areas to processing facilities and are considered non-transportation related activities that are subject to the EPA's spill prevention and response regulations.

C. In Some Cases, PHMSA Has Historically Not Regulated Certain Hazardous Liquid Pipelines for Policy Reasons

Certain exclusions exist to avoid jurisdictional conflicts and the application of duplicative Federal or State regulations. In particular, most producer-operated pipeline facilities on the OCS are regulated by the U.S. Department of the Interior, pursuant to the terms of a Memorandum of Understanding with DOT[19] and non-transportation related facilities, including intra-facility piping, are regulated by the EPApursuant to the Clean Water Act.

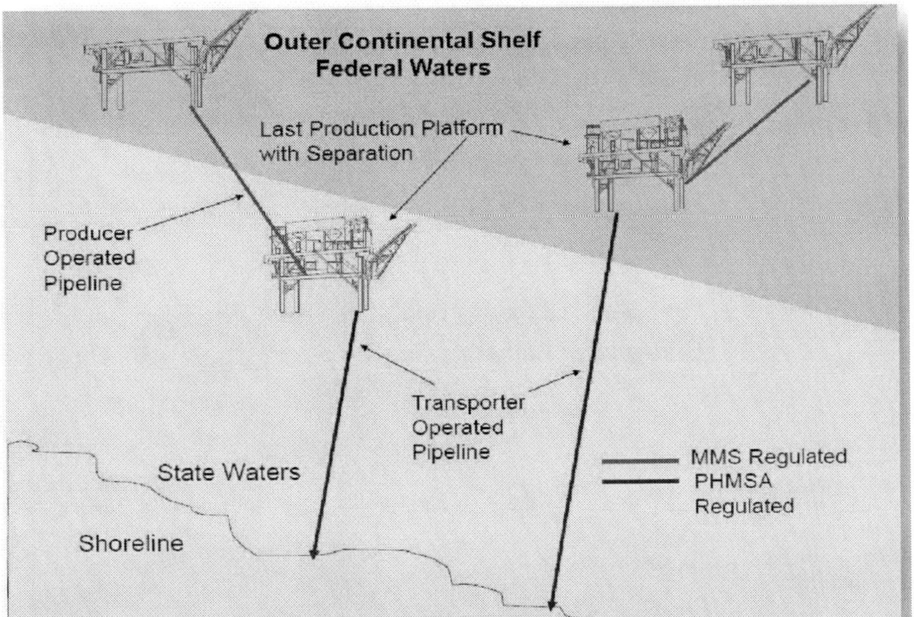

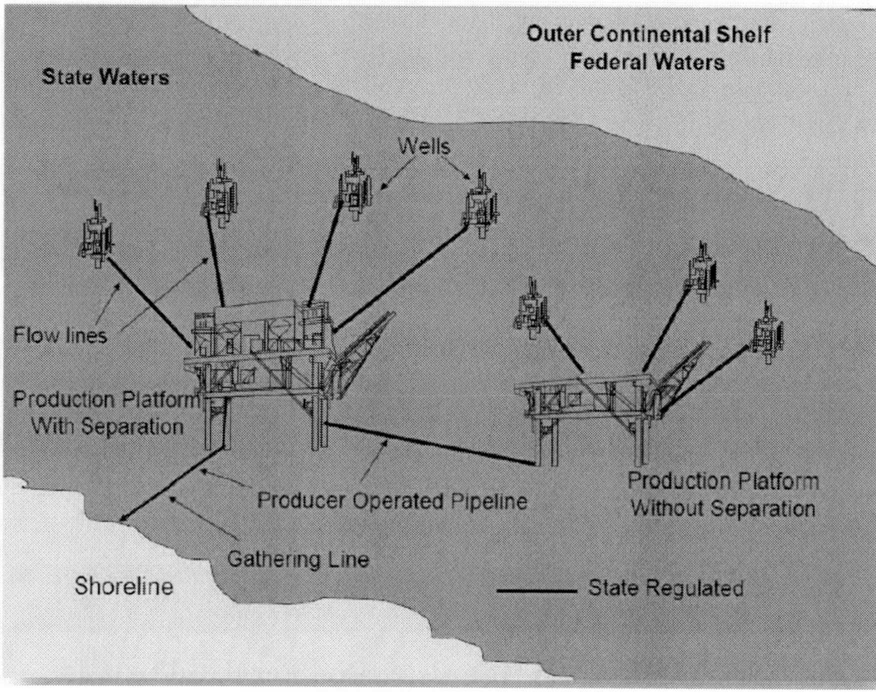

There are also certain offshore pipelines in State waters that are reserved for regulation by the States.[20]

Other exclusions in our regulations exist for historical reasons, including for pipelines that transport hazardous liquids by gravity,[21] or for lack of a sufficient basis for imposing a potential regulation, as in the case of certain small (less than 6") and medium (6" to 8") diameter rural gathering lines.[22]

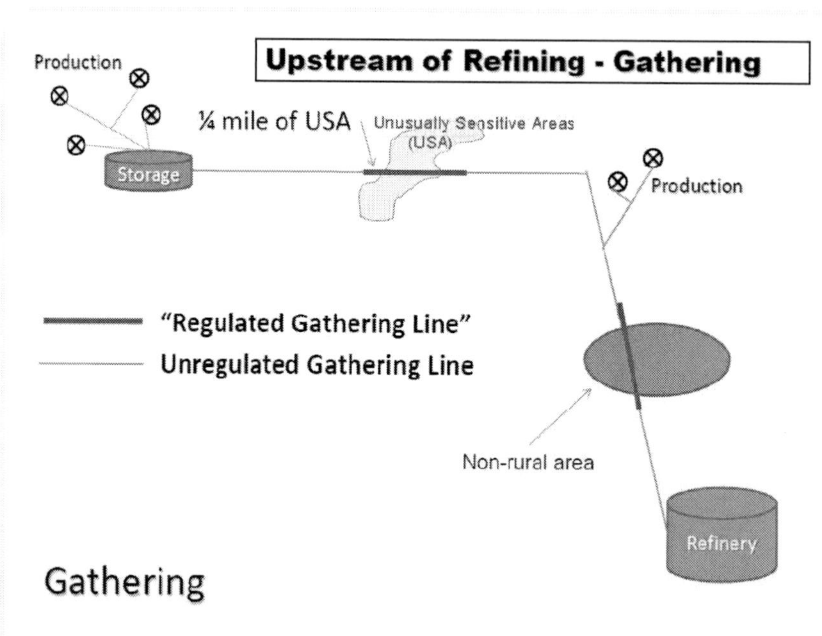

The diagram below illustrates the gathering lines that are currently unregulated:

In summary, PHMSA's statutory and regulatory exclusions are specified in the following provisions:

Unregulated pipelines:	**Statutory Provisions**	**Regulatory Provisions**
Transportation of a hazardous liquid transported in a gaseous state.	60101(a)(4)(B)	49 CFR 195.1(b)(1)
Transportation of a hazardous liquid through a pipeline by gravity.	60101(a)(22)	49 CFR 195.1(b)(2)
A pipeline subject to safety regulations of the U.S. Coast Guard.	60102(k)(3)(A)	49 CFR 195.1(b)(3)
A low-stress pipeline that serves refining, manufacturing, or truck, rail, or vessel terminal facilities, if the pipeline is less than one mile long and does not cross navigable waterway.	60102(k)(3)(B)	49 CFR 195.1(b)(4)
Transportation of hazardous liquid or carbon dioxide in an offshore pipeline in State waters where the pipeline is located upstream from the outlet flange of the following farthest downstream facility.	60101(a)(22)	49 CFR 195.1(b)(5)
Transportation of hazardous liquid or carbon dioxide in a pipeline on the OCS where the pipeline is located upstream of the point at which operating responsibility transfers from a producing operator to a transporting operator.	33 U.S.C. 1520	49 CFR 195.1(b)(6)
A pipeline segment upstream (generally seaward) of the last valve on the last production facility on the OCS where a pipeline on the OCS is producer-operated and crosses into State waters without first connecting to a transporting operator's facility on the OCS.	33 U.S.C. 1520	49 CFR 195.1(b)(7)
Transportation of a hazardous liquid or carbon dioxide through onshore production (including flow lines), refining, or manufac-turing facilities, and associated storage or in-plant piping systems.	60101(a)(22)(B)(ii) and (iii)	49 CFR 195.1(b)(8)
Transportation of a hazardous liquid or carbon dioxide by means other than pipeline.	60101(a)(22)(A)	49 CFR 195.1(b)(9)
Transportation of carbon dioxide downstream from the defined applicable injection points.	60101(a)(22)(B)(ii)	49 CFR 195.1(b)(10)
Not more than 6 inches in diameter, is low pressure, is in rural area, and not unusuallysensitive to environmental damage.	60101(b)(2)(B)	49 CFR 195.1(a)(4)
Less than 6 inches, greater than 20% SMYS, in rural area, and is within ¼ mile of an unusually sensitive area.	60101(b)	49 CFR 195.1(a)(4)
Less than 8 inches, less than 20% SMYS, and is within ¼ mile of an unusually sensitive area.	60101(b)	49 CFR 195.1(a)(4)

III. Improving PHMSA's Statutory Jurisdiction

A. PHMSA is Aggressively, but Appropriately, Administering the Nation's Pipeline Safety Laws

PHMSA has completed nearly all of the mandates and recommendations in the Pipeline Inspection, Protection, Enforcement and Safety Act of 2006. That includes issuing regulations for low-stress pipelines and control room management and completing reports on petroleum market capacity, leak detection technologies, and liquid internal corrosion measures. In addition, PHMSA has sought to improve its relationships with State and local officials and increase its public awareness and outreach program. All States participate in the pipeline safety program, with the exception of Alaska and Hawaii. PHMSA also sponsored the launch of the a nationwide telephonic notice system for damage prevention, the 811 Call-Before-You-Dig Program, and supported the National Association of State Fire Marshals in developing and disseminating training materials for responding to pipeline emergencies.

PHMSA has taken steps to ensure that these efforts continue in the future. For instance, PHMSA has issued a new notice of proposed rulemaking for low-stress pipelines and has plans to bolster its damage prevention program. As indicated in the two diagrams that follow, this closes a significant regulatory gap in pipeline safety:

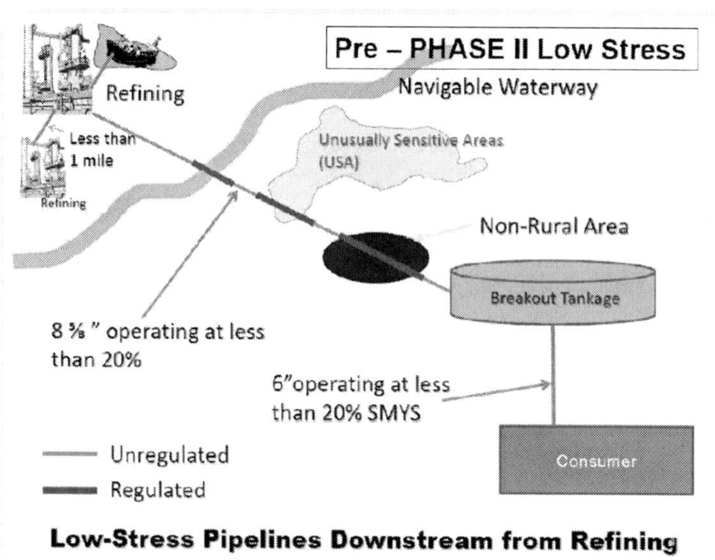

B. PHMSA and Congress Can Work Together to Improve the Regulation of Transporting Hazardous Liquids by Pipeline

The support of Congress is critical to the safe and effective regulation of the transportation of hazardous liquids by pipeline. PHMSA is in the process of developing legislation that would address our jurisdiction over the transportation of hazardous liquids by pipeline in the future. We look forward to working with Congress to address any issues you may have concerning PHMSA's pipeline safety program and the regulation of hazardous liquid pipelines. PHMSA very much appreciates the opportunity to report on our authority over these pipelines and the opportunities that exist to strengthen oversight.

Thank you. I would be pleased to answer any questions you may have.

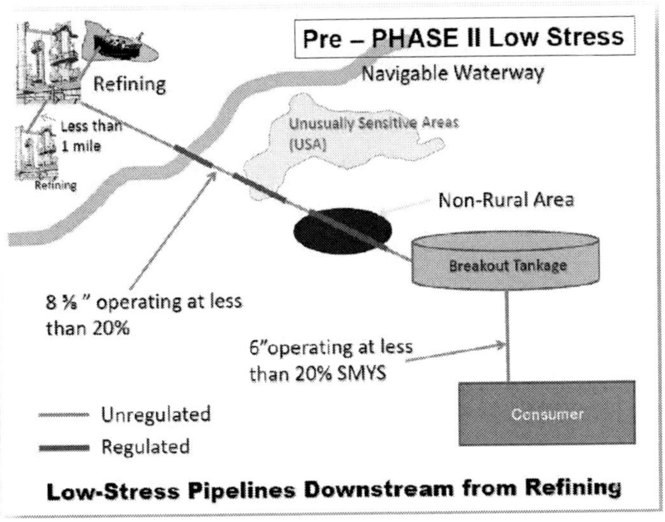

End Notes

[1] The Pipeline Safety Act (PSA), 49 U.S.C.A. § § 60101 et seq., enacted in 1994, combined and recodified, without substantive changes, the two then existing pipeline safety statutes, the Hazardous Liquid Pipeline Safety Act of 1979 (former 49 U.S.C.A. § § 2001 to 2014) (HLPSA) and the Natural Gas Pipeline Safety Act of 1968 (former 49 U.S.C.A. § § 1671 et seq.) (NGPSA).
[2] Gonzales v. Raich, 545 U.S. 1, 16-17 (2005).
[3] *Id.* § 60101(a)(4).
[4] *Id.* § 60101(a)(22).
[5] *Id.* § 60101(b)(2).
[6] *Id.* § 60102(k).
[7] *See id.* § 60101(a)(22).
[8] *See* 40 C.F.R. § 112
[9] 49 U.S.C. § 60104(e).
[10] 49 U.S.C. § 60102(b)(2).
[11] 49 U.S.C. § 60102(b)(5).
[12] Pub. L. No. 91-190, 83 Stat. 852 (1970) (amended by Pub. L. No. 94-52 (1975); Pub. L. No. 94-83 (1975); Pub. L. No. 97-258, § 4(b) (1982); currently codified at 42 U.S.C. §§ 4321-4347).
[13] 44 U.S.C. §§ 3501-3520.
[14] Pub. L. No. 104-113, 110 Stat. 775 (1996).
[15] 49 U.S.C. § 60105 (2006).
[16] *Id.* § 60106(b)(1).
[17] *Id.* § 60104(c).
[18] *See id.* § 60101(a)(22).
[19] *See* 49 C.F.R. § 195.1(b)(6)-(7) (2010); Pipeline Safety: Regulations Implementing Memorandum of Understanding With the Department of the Interior, 62 Fed. Reg. 61692 (Nov. 19, 1997); 33 U.S.C. § 1520; *see also* Memorandum of Understanding Among the Secretary of the Interior, Secretary of Transportation, and Administrator of the Environmental Protection Agency (1994); Oil Pollution Prevention; Non-Transportation-Related Onshore Facilities, 59 Fed. Reg. 34070 (July 1, 1994).
[20] 49 C.F.R. § 195.1(b)(5).
[21] Explosives and Other Dangerous Articles: Pipeline Transportation, 32 Fed. Reg. 1098 (Jan. 31, 1967).
[22] *See* 49 U.S.C. §60101(b)(2)(B)(ii).

In: Pipeline Safety and Security
Editor: Jonathan T. Miller and Lisa M. Gurton
ISBN: 978-1-61209-147-1

Chapter 7

TESTIMONY OF THE HONORABLE DEBORAH A.P. HERSMAN, CHAIRMAN, NATIONAL TRANSPORTATION SAFETY BOARD, BEFORE THE SUBCOMMITTEE ON RAILROADS, PIPELINES, AND HAZARDOUS MATERIALS, HEARING ON "THE SAFETY OF HAZARDOUS LIQUID PIPELINES: REGULATED VS. UNREGULATED PIPELINES"

Deborah A.P. Hersman

INTRODUCTION/OVERVIEW

Chairman Brown, Ranking Member Shuster, Members of the Subcommittee, thank you for the opportunity to address you today on the safety of hazardous liquid pipelines. According to the Pipeline and Hazardous Materials Safety Administration (PHMSA), there are approximately 173,000 miles of hazardous liquid pipelines. The National Transportation Safety Board (NTSB) plays an important role in promoting the safe transit of liquid and gas materials through this pipeline system by investigating accidents and issuing safety recommendations.

While PHMSA has met several of the NTSB's recommendations to improve pipeline safety, we remain concerned about certain aspects of PHMSA's pipeline safety program. Two such areas specifically addressed in the Pipeline, Inspection Protection Enforcement Safety (PIPES) Act of 2006 are the regulation of low-stress pipeline systems and requirements for the use of excess flow valves.

REGULATION OF LOW-STRESS PIPELINE SYSTEMS

Corrosion failures on the BP Exploration, Inc.'s, low-stress oil transit lines from the Prudhoe Bay oil fields to the Trans Alaska pipeline in 2006 raised concerns among Members of Congress about the potential pollution of environmentally sensitive areas. As a result, Congress included provisions in the PIPES Act mandating that PHMSA issue regulations subjecting low- stress hazardous liquid pipelines near unusually sensitive environmental areas to the same standards and regulations as other hazardous liquid pipelines. Low-stress pipelines are those that are operated at a stress level of 20 percent or less of their strength ratings.

At the time the PIPES Act was enacted, federal pipeline safety regulations only applied to low-stress pipelines that were located in populated areas, crossed navigable waterways, or carried highly volatile liquids, such as compressed liquefied propane. In a final rulemaking, "Pipeline Safety: Protecting Unusually Sensitive Areas from Rural Onshore Hazardous Liquid Gathering Lines and Low-Stress Lines," published on June 3, 2008, PHMSA issued regulations for rural onshore low-stress pipelines that have a diameter of at least 8 5/8 inches and that are within 1/2 mile of an area defined as unusually sensitive. Low-stress pipelines meeting these criteria will be required to meet Title 49 Code of Federal Regulations, Part 195, for hazardous liquid pipelines in its entirety by July 2012.

The final rule also included regulations for rural onshore gathering lines that operate at stress levels greater than 20 percent of the pipe strength, have a diameter between 6 5/8 and 8 5/8 inches and are within 1/4 mile of an area defined as unusually sensitive. A "gathering line" is defined as a pipeline with a diameter of 8 5/8 inches or less that transports petroleum from a production facility. Under the final rule, rural onshore gathering lines will be required to meet Part 195 in part by July 2011. The safety requirements of Part 195 that will eventually apply to the rural onshore gathering lines include annual and accident reporting requirements, establishment of maximum operating pressure, installation of line markers, public education programs, damage prevention programs, corrosion control, and operator qualification programs.

On June 22, 2010, PHMSA published a follow-up Notice of Proposed Rulemaking (NPRM) addressing the regulation of all rural onshore hazardous liquid low-stress pipelines. This NPRM represents phase two of PHMSA's implementation of its mandate in the PIPES Act. In this NPRM, PHMSA proposes safety requirements for all rural low-stress pipelines not included under the phase one final rule. This latest NPRM does not include any new proposed requirements for onshore rural gathering lines.

The low-stress pipelines captured under the new NPRM include (1) rural low-stress pipelines of a diameter less than 8 5/8 inches located in or within one-half mile of an unusually sensitive area and (2) all other rural low-stress pipelines that were not included under phase one. PHMSA estimates that the NPRM will apply to 1,384 miles of low-stress pipelines not covered by the previous rule. Under the new NPRM, PHMSA proposes to establish three categories of rural onshore low-stress pipelines that delineate decreasing levels of risk. Category 1 includes those low-stress pipelines covered in the June 2008 rulemaking. Categories 2 and 3 include low- stress pipelines of decreasing risk on the basis of pipeline size and location. PHMSA is proposing partial compliance with Part 195 for category 2 and 3 low-stress pipelines.

The NTSB believes PHMSA has taken a reasonable approach to the regulation of hazardous liquid low-stress pipelines. However, as I will mention momentarily, the key to the success of these regulations will be effective oversight exercised by the pipeline operators and PHMSA.

The NTSB would also like to note the regulation of offshore pipeline systems has not been addressed in recent legislation or regulatory action. The NTSB recognizes that jurisdiction of offshore pipelines of all types is complex and involves the states, PHMSA, and the Department of the Interior.

The tragedy in the Gulf of Mexico involving the Deepwater Horizon drilling platform is a grim reminder of the damage that a major oil spill can cause. While the magnitude of the Deepwater Horizon spill is far greater than any known pipeline failure, the events in the Gulf should remind those involved in the pipeline industry that all pipelines must be sufficiently safeguarded and regulated in order to protect the public and the environment.

Integrity Management Programs for Distribution Systems and the Use of Excess Flow Valves

The PIPES Act also mandates that DOT prescribe minimum standards for integrity management programs for distribution pipeline systems. On June 25, 2008, PHMSA published an NPRM, "Integrity Management Program for Gas Distribution Pipelines", with proposed regulations that would require operators of gas distribution pipelines to develop and implement integrity management programs with the same objectives as the existing integrity management programs for hazardous liquid and gas transmission pipelines.

Integrity management programs for hazardous liquid and gas transmission pipelines typically require operators to assess the condition of their pipelines by using "in-line" inspection tools that travel through the pipeline to determine the nature and extent of any defects or pressure testing that yields information about the integrity of the pipeline. Such techniques are not feasible for typical distribution pipeline systems because of the differences in the design and operating parameters between distribution pipeline systems and hazardous liquid and gas transmission pipelines.

Further, the failure of a distribution pipeline is often initially detected from reports of a gas leak rather than a catastrophic rupture. As result, development and implementation of an effective leak management program is an important element of an integrity management program for a distribution pipeline.

PHMSA acknowledged these differences in the NPRM and properly emphasized the importance of various leak detection methods as essential elements of an integrity management program for distribution pipeline systems.

In its comments on the NPRM, the NTSB emphasized that while an effective leak detection program is a crucial element of the overall leak management program, the use of equipment that prevents or mitigates leaks is equally important. One such device that mitigates a gas pipeline leak is an "excess flow valve." An excess flow valve is a device installed on the distribution line, usually serving a user residence or facility, that detects an abnormally high flow rate, and when an excess flow is detected, automatically closes a valve, thus shutting off the flow of gas through the distribution line. The NPRM did not adequately address this aspect of leak management, other than incorporating the mandate for PHMSA to require excess flow valves on new or replacement distribution lines serving single family residences. PHMSA complied with this provision of the PIPES Act on December 4, 2009, when it published the final rule on integrity management programs for distribution pipeline systems.

The NTSB has long advocated the use of excess flow valves in gas distribution pipeline systems as an effective means of preventing explosions caused by natural gas leaking from distribution systems. On July 7, 1998, a natural gas explosion and fire destroyed a newly constructed residence in South Riding, Virginia, a suburb of Washington. The accident caused one fatality and one serious injury. The NTSB determined that the gas service line to the home had failed and that an uncontrolled release of gas had accumulated in the basement and subsequently ignited. The NTSB concluded from its investigation that had an excess flow valve been installed in the service line, the valve would have closed shortly after the hole in the service line developed and the explosion likely would not have occurred. The NTSB recommended that PHMSA require excess flow valves be installed in all new and renewed gas service lines, regardless of a customer's classification, when the operating conditions are compatible with readily available valves. The NTSB believes that apartment buildings, other multifamily dwellings, and commercial properties are susceptible to the same risks from leaking gas lines as single-family residences, and we believe this gap in the law and the regulations should be eliminated.

Oversight of Integrity Management and Other Risk-Based Pipeline Safety Programs

Over the past decade or more, PHMSA has used a risk-based assessment for regulating the DOT pipeline safety program. PHMSA has successfully built a partnership with various facets of the pipeline industry to develop, implement and execute a multi-part pipeline safety program. In the NTSB's view, all stakeholders, including PHMSA, have come to rely heavily upon this approach. The NTSB believes that a risk-based methodology can be effective in developing and executing pipeline safety programs, and there are many positive elements to PHMSA's approach.

The DOT pipeline safety regulations based on risk assessment principles provide the structure, content, and scope for many aspects of the overall pipeline safety program. Within this regulatory framework, pipeline operators have the flexibility and responsibility to develop their individual programs and plans, determine the specific performance standards, implement their plans and programs, and conduct periodic self-evaluations that best fit their particular pipeline systems. PHMSA likewise has the responsibility to review pipeline operators' plans and programs for regulatory compliance and effectiveness.

The NTSB believes that with the risk-based assessment come increased responsibilities for both the individual pipeline operators and PHMSA. Operators must diligently and objectively scrutinize the effectiveness of their programs, identify areas for improvement, and implement corrective measures. PHMSA, as the regulator, must also do the same in its audits of the operators' programs and in self-assessments of its own programs. In short, both operator and regulator need to verify whether risk-based assessments are being executed as planned, and more importantly, whether these programs are effective.

Following pipeline accidents in Mounds View, MN and Edison, NJ, the Board issued recommendations to PHMSA to expedite requirements for the installation of remotely operated valves. Our recommendations were eventually closed in 2004 based on PHMSA's assertion that integrity management rules for high consequence areas would require operator evaluation of the need for and installation of emergency control devices, like remotely operated valves. NTSB has supported the integrity management process in principle, but also believes the critical component of successful integrity management depends upon the diligence of each individual operator, PHMSA, and its delegated State enforcement agencies.

Unfortunately, there have been some recent pipeline investigations in which the NTSB discovered indications that PHMSA and operator oversight of risk-based assessment programs, specifically integrity management programs and public education programs, has been lacking and has failed to detect flaws and weaknesses in such programs.

In its investigation of the October 2004, rupture of an anhydrous ammonia pipeline near Kingman, Kansas, the NTSB identified deficiencies in PHMSA's auditing procedures when evaluating the operator's integrity management program. The operator did not include assessments of leak history when calculating relative risk scores for various segments of the pipeline. These relative risk scores were used to establish an initial baseline assessment of the integrity of the pipeline in the decision-making process for prioritizing the inspection schedule. Though PHMSA did find omissions of other risk factors during its review of the operator's integrity management program, PHMSA did not identify the omission of the leak history data during its initial review or during a subsequent review of the corrected plan. Consequently, the ruptured pipeline segment was not scheduled for a baseline assessment until 2006, almost 2 years after the October 27, 2004, rupture. The NTSB recommended that PHMSA require an operator to revise its pipeline risk assessment plan whenever it has failed to consider one or more risk factors that can affect pipeline integrity.

The November 1, 2007, rupture of a propane pipeline in Carmichael, Mississippi, resulted in two fatalities, seven injuries, and property damage exceeding $3 million. Before the accident, the pipeline operator relied upon contractors to obtain accurate mailing data and ensure that mailings to the public were completed. However, the operator did not perform oversight to ensure that all appropriate recipients were on the mailing lists and that the mailings met appropriate regulatory requirements. The operator also had not taken any action to determine whether recipients who received the mailings understood the guidance they contained. The NTSB determined that the pipeline operator failed to properly assess its public awareness and education program by relying upon contractors without appropriate oversight. The NTSB recommended that PHMSA initiate a program to evaluate pipeline operators' public education programs, including the operators' self-evaluations of the effectiveness of their public education programs.

On May 4, 2009, an 18-inch diameter gas transmission pipeline with an operating pressure of 850 psi ruptured near Palm City, Florida. The rupture was located in the Florida Turnpike right-of-way, between I-95 and the Florida Turnpike. The turnpike and interstate were closed for approximately three hours due to the accident. Two gas transmission pipelines operated by the same pipeline company were also located in the right-of-way but were reportedly not damaged.

The force of the released gas created a crater approximately 116.5 feet long by 17 feet wide by approximately 2.8 feet deep. Roughly 104 feet of the pipe was ejected from the ruptured pipeline and landed next to the crater. The closest edge of the crater was approximately 25 feet from the northbound paved edge of the Florida Turnpike.

There was no ignition of the released gas, and no fatalities were reported. However, two people were injured when their car reportedly hit debris, ran off the road, and turned over; a deputy sheriff was hospitalized after walking through a gas cloud; and the accident resulted in the evacuation of a nearby school and residential community.

The NTSB's ongoing investigation has determined that at the time of the accident, the operator had not identified the ruptured segment as located within a high consequence area, and therefore the segment was not covered by the operator's integrity management plan. However, an independent evaluation done by PHMSA at the NTSB's request shows the segment, in fact, is in a high consequence area. The NTSB is collecting documentation to determine the cause of this error.

As a result of these investigations, the NTSB is concerned that the level of self-evaluation and oversight currently being exercised is not uniformly applied by some pipeline operators and PHMSA to ensure that the risk-based safety programs are effective. The NTSB believes that to ensure effective risk-based integrity management programs are employed throughout the pipeline industry, PHMSA must establish an aggressive oversight program that thoroughly examines each operator's decision-making process for each element of its integrity management program.

Recent Accidents in Texas

The two most recent pipeline accidents in Cleburne, Texas and Darrouzett, Texas involved third-party excavation damage resulting in ruptures, fires, and explosions. Preliminary information from both investigations indicates that prior to the start of excavation activities, neither pipeline was marked or identified. Both investigations will determine the reasons why and how these lapses occurred.

Cleburne, TX Summary

On June 7, 2010, a natural gas transmission pipeline measuring 36-inches in diameter near Cleburne, Texas was struck and ruptured by a contractor for an electrical cooperative that was installing a pole for a power line. One member of the contractor's crew was drilling a hole while operating an auger affixed to a truck when the auger struck and punctured the transmission pipeline. An ignition and explosion of the escaping gas resulted, and the operator of the auger was killed. Six other crewmen were hospitalized.

The accident pipeline had a nominal wall thickness of 0.5-inch. The pipeline was operating at 950 psi at the time of the accident. The maximum allowable operating pressure is 1,050 psi. The pipeline, constructed in 1971, is 388 miles long, originating in Coyanosa, Texas and terminating in Ennis, Texas.

A second pipeline operated by a different pipeline company also traversed the accident area. Workmen in the area reported that they saw markers for the second pipeline. An NTSB investigator and Texas Railroad Commission personnel visiting the site also observed markers for the second pipeline, but the ruptured pipeline was not marked.

The NTSB is currently investigating this accident with the assistance of PHMSA and the Texas Railroad Commission (the state regulatory agency for pipeline safety).

Darrouzett, TX Summary

On June 8, 2010, a natural gas non-regulated gathering line measuring 14-inches was struck by a third party contractor near Darrouzett, Texas. The maximum allowable operating pressure of the gathering line was 700 psi; the line was operating at approximately 500 psi. The line begins in Follett, Texas, travels into Oklahoma, continues west and then returns to Texas near the Hansford/Sherman County area. The line is fed by many gathering lines in the area and ends at the plant in Sherman, Texas.

At the time of the incident, six contractor personnel were working in the area. Two persons were killed, one critically injured, and three others escaped injury. A bulldozer working in a caliche pit struck the 14-inch natural gas pipeline sometime before 4pm. The pipeline operator's SCADA system picked up a pressure loss and began closing valves to isolate the ruptured section of the pipeline. The fire was extinguished by 8 pm.

Preliminary information from the Texas Railroad Commission, the lead agency in this investigation, indicates that the excavator had not requested a permit to work in the area nor were there any pipeline markers at the accident scene. The accident gathering line is not regulated under DOT pipeline regulations.

PHMSA accident statistics over the past decade (2000-2009) identify corrosion as the leading cause of all reported pipeline accidents. The second leading reported cause is damage from third party excavators. Despite the focus on one-call systems, marking of pipelines prior to excavation, and other measures, the two accidents in Texas are a reminder that excavation damage remains a serious concern.

CLOSING

In summary, the NTSB believes more can be done to improve pipeline safety, and thus the safety of people living and working near, and receiving service from, our nation's pipelines.

This concludes my testimony and I would be happy to answer any questions you may have.

In: Pipeline Safety and Security
Editor: Jonathan T. Miller and Lisa M. Gurton

ISBN: 978-1-61209-147-1

Chapter 8

ACTIONS NEEDED TO ENHANCE PIPELINE SECURITY

Pipeline and Hazardous Materials Safety Administration

Subject: *ACTION:* Actions Needed To Enhance Pipeline Security
Pipeline and Hazardous Materials Safety Administration
Report Number AV-2008-053

Date: May 21, 2008

From: Lou E. Dixon
Assistant Inspector General
for Aviation and Special Program Audits

Reply to Attn. of: JA-10

To: Pipeline and Hazardous Materials Safety Administrator

This report presents the results of our review on multi-agency actions to implement the pipeline security annex to the Memorandum of Understanding (MOU) between the Department of Transportation (DOT) and Department of Homeland Security (DHS). The Pipeline Inspection, Protection, Enforcement, and Safety Act of 2006[1] requires the Office of Inspector General to assess actions taken to implement the annex.

Consistent with the act, our audit objectives were to: (1) assess the adequacy and effectiveness of the process by which DOT communicates and coordinates with DHS on matters of pipeline security; (2) assess the agencies' status in implementing the program elements outlined in the annex; and (3) review the role, responsibility, and authority of DOT regarding pipeline security matters. We conducted this review between June and November of 2007 in accordance with generally accepted *Government Auditing Standards* as prescribed by the Comptroller General of the United States. Exhibit A details our objectives, scope and methodology, and related audits. Exhibit B outlines the program elements in the annex.

SECURING THE NATION'S PIPELINE SYSTEMS IS A CONTINUING CHALLENGE

The Nation's pipeline infrastructure is a network of approximately 2 million miles of pipelines that move millions of gallons of hazardous liquids and billions of cubic feet of natural gas daily. Within the United States, there are about 2,200 natural gas pipeline operators and 300 hazardous liquids pipeline operators.

DOT's Pipeline and Hazardous Materials Safety Administration (PHMSA) is responsible for overseeing the safety of the Nation's pipeline system. The Transportation Security Administration (TSA) within DHS is responsible for securing the Nation's transportation infrastructure, including pipelines.

Safeguarding the Nation's massive pipeline infrastructure from catastrophic events (i.e., terrorism or natural disasters) is a continuing challenge for DOT and DHS. In 2005, Hurricane Katrina devastated the Gulf Coast region and demonstrated the vulnerabilities of the Nation's critical infrastructure. Loss of electrical power to pumping stations forced three major pipeline operators to shut down. This eliminated most fuel sources to the entire eastern seaboard and caused a vast array of economic disruptions, including hoarding and severe price spikes.

It should be noted that the most frequently targeted mode of transportation by terrorists worldwide is pipeline systems. In Colombia, for example, rebels have bombed the Caño Limón oil pipeline over 600 times since 1995. Terrorist plots against pipelines have also occurred within the United States.

- In December 1999, Federal agents arrested two anti-government militia members for plotting to detonate 24 million gallons of liquid propane at a storage facility in Elk Grove, California.
- In June 2007, the U.S. Department of Justice arrested members of a terrorist group planning to attack jet fuel pipelines and storage tanks at the John F. Kennedy (JFK) International Airport in New York.
- In November 2007, a U.S. citizen was sentenced to 30 years in Federal prison for plotting to help an alleged al-Qaeda operative blow up U.S. oil pipelines and refineries.

In addition, pipeline incidents can have deadly implications, such as the August 19, 2000, natural gas transmission pipeline (30-inch-diameter) that ruptured adjacent to the Pecos River near Carlsbad, New Mexico. The released gas ignited and burned for 55 minutes. Twelve people who were camping under a concrete-decked steel bridge that supported the pipeline across the river were killed, and their three vehicles were destroyed. Two nearby steel suspension bridges for gas pipelines crossing the river were also extensively damaged.

These events underscore the need for a well-defined, well-coordinated, interagency approach to prevent, detect, and respond to such events. As we have stated in our DOT Top Management Challenges reports,[2] there is growing interdependency between DOT and other Federal agencies to ensure safe, secure, and efficient transportation. This is most evident in the responsibility to protect the public from criminal and terrorist acts, particularly in DOT's working relationship with DHS.

In today's constrained fiscal environment, DOT and DHS must leverage their resources to secure the Nation's pipelines. TSA's pipeline security program— with just 11 personnel— has the biggest challenge to effectively oversee the vast network of natural gas and hazardous liquids pipeline operators. Although PHMSA has regional offices and about 80 inspectors nationwide, it partners with state agencies—which have over 400 inspectors—to oversee and enforce compliance with pipeline safety requirements, primarily operators of natural gas distribution pipeline systems. Thus, it is incumbent upon PHMSA, TSA, and their state partners to effectively work together to enhance the security of the Nation's pipeline infrastructure.

Congress continues to emphasize the importance of securing the Nation's pipelines and related infrastructure. On August 3, 2007, the President signed into law the Implementing Recommendations of the 9/11 Commission Act of 2007 (9/11 Commission Act).[3] The law mandates, by 1 year after enactment, that the Secretary of Homeland Security and the Secretary of Transportation develop and implement a plan to review and inspect the 100 most critical pipeline operators' pipeline security plans and critical facilities.

RESULTS IN BRIEF

We found that PHMSA and TSA have taken initial steps toward formulating an action plan to implement the provisions of the pipeline security annex, which are important to enhance pipeline security, clearly define roles, maximize resources, and prevent duplication of effort. However, further actions need to be taken with a sense of urgency because the current situation is far from an "end state" for enhancing the security of the Nation's pipelines.

In September 2004, DOT and DHS entered into an MOU to facilitate the development and deployment of transportation security measures. In August 2006, PHMSA and TSA signed an annex to the MOU to establish clear lines of authority and responsibility over pipeline security matters. To accomplish this, the annex requires PHMSA and TSA to jointly develop an action plan by February 2007 to implement the annex provisions and program elements. Some program elements of the annex include identifying critical infrastructure and key resources; performing risk assessments; strategic planning; developing regulations, guidelines and directives; and conducting inspection and enforcement actions.

PHMSA and TSA established a joint working group tasked with developing a multi-year action plan to execute the annex provisions and program elements. However, PHMSA and TSA have neither finalized the action plan to implement the annex nor initiated key program elements due to a working relationship that was not always cooperative. Both PHMSA and TSA agreed that they did not always coordinate their efforts in a collegial manner.

Beginning in the second half of 2007, PHMSA and TSA began improving their coordination to address the annex provisions and program elements. For example, PHMSA created and staffed a new management position within the Office of the PHMSA Administrator. This manager is responsible for all aspects of PHMSA's emergency management and security policy activities and reports directly to the Chief Safety Officer. This individual will also liaison with other agencies, including TSA and other stakeholders, regarding emergency management and security policy issues.

To further enhance the security of the Nation's pipeline infrastructure, PHMSA and TSA need to complete the following actions:

Finalize and effectively execute the action plan to implement the security annex. PHMSA and TSA signed the annex 21 months ago, but the two agencies still have not finalized the action plan to fully implement the program elements of the annex—even though the deadline (February 2007) elapsed over a year ago. In December 2007, the agencies had not completed 9 of the 11 program elements, and the action plan did not contain several initiatives designed specifically to enhance coordination efforts. These include procedures to coordinate on observations or recommended measures from vulnerability assessments to determine if they conflict with current or planned safety requirements. Interagency coordination is essential to prevent the implementation of security recommendations that could inadvertently contradict safety regulations and put the safety of the Nation's pipelines at risk.

During the first 2 months of 2008, the two agencies began addressing outstanding annex program elements. PHMSA and TSA developed a new action plan, dated February 11, 2008, which estimates that most of the program element initiatives will be complete by the last quarter of 2009. While PHMSA and TSA have made progress toward finalizing the action plan, this progress began nearly a year after the completion deadline agreed to in the annex. Both agencies must sustain and expedite these collaborative efforts to finalize and effectively execute the annex provisions and program elements.

Delineate roles and responsibilities in overseeing and enforcing security regulations for operators of liquid natural gas (LNG) facilities.[4] By law, TSA holds the lead authority and primary responsibility for security activities in all modes of transportation, including pipelines. TSA currently conducts reviews of pipeline operators' compliance with TSA voluntary guidance[5] but does not have regulations related to pipeline security and does not take enforcement actions against pipeline operators. PHMSA has—and enforces—its own security regulations that existed prior to the creation of TSA in 2001. As a result, both agencies conduct reviews of LNG operators' security procedures. PHMSA's regulations have specific security requirements[6] for operators of LNG facilities, and PHMSA inspects LNG facilities to ensure that they meet those requirements. TSA has stated that it can issue security directives, but it has not done so. These directives would allow TSA to take enforcement actions against pipeline operators.

Because the annex does not explicitly state which agency has primary oversight and enforcement authority for LNG operators, there is a lack of clearly defined roles at the working level. This could cause pipeline operators to receive conflicting or duplicative guidance, which will ultimately impede security measures. To resolve this issue, the annex should be amended to clearly delineate the roles and responsibilities of PHMSA and TSA in overseeing and enforcing security regulations for LNG operators.

Maximize the strategy to assess pipeline operators' security plans and guidance. The 9/11 Commission Act requires DOT and DHS to develop a plan to review the 100 most critical operators' security plans and critical facilities by August 2008.[7] The act also requires that DOT and DHS issue security recommendations by February 2009 to operators of natural gas and hazardous liquids pipelines and pipeline facilities. The act also stipulates that if DHS determines that regulations are appropriate, it is required to consult with DOT. One of the two agencies shall then promulgate the regulations and perform necessary inspection and enforcement functions. The need for new security regulations will be partly determined by the degree to which pipeline operators are following TSA's current security guidance.

However, TSA's current security guidance is not mandatory and remains unenforceable unless a regulation is issued to require industry compliance. To adequately determine if new security regulations are needed, PHMSA and TSA will need to conduct covert tests of pipeline systems' vulnerabilities to assess the current guidance as well as evaluate operators' compliance. Until this is done, there is no way to effectively assess the adequacy of the security guidance or to measure operator compliance with the guidance.

In our opinion, actions are needed from both PHMSA and TSA to maximize resources for assessing pipeline operators' security plans and guidance. These actions should include:

- Ensuring that PHMSA is actively involved in the inspections of the 100 most critical operators' security plans and critical facilities. To date, PHMSA's role has been limited to an "as needed" basis. According to PHMSA, it has not regularly attended past TSA security reviews of pipeline operators. To ensure effective and timely execution of this congressional mandate, PHMSA should participate in these inspections on a regular basis, especially given its level of expertise in security-related matters.
- Developing testing protocols and performing vulnerability tests to ascertain, among other things, if unauthorized individuals can penetrate operators' critical infrastructure (including cyber attacks). Currently, there are no plans to develop protocols and conduct vulnerability tests. Without testing, there is no way to effectively measure operators' compliance with existing guidance or assess the adequacy of the guidance.

PHMSA and TSA must work together to develop a pipeline security strategy that maximizes the value and efficiency of both agencies' efforts. This is a fundamental factor in enhancing pipeline security. PHMSA must make every effort possible to communicate and coordinate with TSA to resolve the issues identified in this report. If PHMSA and TSA are unable to expedite the actions needed and improve coordination, it may be necessary for Congress to take action.

Our complete recommendations are listed on page 12. PHMSA's comments and our response are discussed on pages 13 and 14.

FINDINGS

PHMSA and TSA have taken initial steps toward formulating an action plan to implement the provisions of the annex; however, further actions are needed as the current situation is far from an "end state" for enhancing the security of the Nation's pipeline system.

After PHMSA and TSA signed the annex, they designated a joint working group to develop a multi-year action plan for implementing the provisions and program elements of the annex. The working group was to complete its efforts on developing the action plan by February 2007. Almost a year later, the working group developed a multi-year action plan, which identified many, but not all, of the initiatives and milestones to execute each program element.

Based on our review, PHMSA and TSA need to complete the following actions to further enhance the security of the Nation's pipeline infrastructure: (1) finalize and effectively execute the action plan for implementing the security annex, (2) amend the annex to delineate the roles and responsibilities of PHMSA and TSA in overseeing and enforcing security regulations for LNG operators, and (3) maximize the strategy used to assess pipeline operators' security plans and guidance to ensure effective and timely execution of congressional mandates in the 9/11 Commission Act.

PHMSA and TSA Need to Finalize and Effectively Execute the Action Plan for Implementing the Security Annex

PHMSA and TSA signed the annex 21 months ago; yet, the two agencies have neither finalized the action plan to implement the annex program elements nor developed several initiatives. This has delayed key elements of the annex that were designed specifically to enhance communication and coordination efforts between the two agencies. For example, PHMSA and TSA have not developed an initiative to coordinate on observations or recommended measures from vulnerability assessments to determine if they conflict with current or planned safety requirements. This coordination is essential to prevent security recommendations that could unintentionally contradict safety regulations and put the safety of the Nation's pipelines at risk.

Towards the end of 2007, we found that the agencies had not completed 9 of the 11 annex program elements because there are no deadlines to foster timely reviews and decision processes; further, the agencies are not held accountable for failure to abide by the annex provisions. Without interim deadlines and accountability, there is no guarantee the action plan will be finalized and properly executed.

For example:

- The December 2007 action plan to implement the annex did not contain some of the initiatives for program elements. For instance, the plan did not include certain items that the agencies agreed to in the annex, such as (1) an initiative for PHMSA to provide TSA with data and information collected during PHMSA's security inspections or reviews of security plans, (2) protocols for ongoing information sharing and participation in their respective research and development planning, and (3) an initiative for TSA to coordinate with PHMSA on observations or recommended measures—derived from the results of criticality and vulnerability assessments—to evaluate whether they conflict with or adversely affect current or planned safety requirements.

- PHMSA and TSA had not established target dates to complete some of the elements and element initiatives. Under the annex, PHMSA and TSA agreed to develop a plan with specific timeframes for implementing the program elements. For example, the December 2007 plan did not contain timeframes to: (1) develop a procedure for requesting special permits to install pipeline facilities in the event of a security incident, (2) provide training to TSA staff on technical issues related to PHMSA's mission, or (3) perform a study on the petroleum pipeline network supply.
- Some elements have future target dates that extend out to 2009 (we are concerned that PHMSA and TSA will not meet these deadlines, since they have been unable to finalize the action plan to execute the annex).

During the first 2 months of 2008, the two agencies began addressing outstanding annex program elements. PHMSA and TSA developed a new action plan, dated February 11, 2008, which estimates that most of the program element initiatives will be complete by the last quarter of 2009. Also, the plan now contains initiatives for PHMSA to invite TSA to its security reviews of LNG facilities and for PHMSA and TSA to share security-related enforcement actions. However, the current action plan still does not contain initiatives (1) to develop protocols for ongoing information sharing and participation in the agencies' respective research and development planning and (2) for TSA to coordinate with PHMSA on observations or recommended measures—derived from the results of criticality and vulnerability assessments—to evaluate whether they conflict with or adversely affect current or planned safety requirements.

While PHMSA and TSA have made progress toward finalizing the action plan, this progress began nearly a year after the completion deadline agreed to in the annex. Both agencies must sustain and expedite these collaborative efforts to finalize and effectively execute the annex provisions and program elements.

PHMSA and TSA Need To Delineate Their Roles and Responsibilities for Overseeing and Enforcing Security Regulations for Operators of LNG Facilities

A central goal of the annex was to delineate clear lines of authority and prevent duplication of effort. Yet, the annex does not contain language that explicitly states which agency will be responsible overseeing and enforcing security regulations for operators of LNG facilities. Since both PHMSA and TSA can conduct reviews of LNG facilities, a clear line of authority does not exist. This creates the potential for duplicative efforts and confusion among LNG operators regarding which agency they should look to for guidance as the lead Federal security regulator.

By law, TSA holds the lead authority and primary responsibility for security activities in pipelines. In practice, PHMSA's role, responsibility, and authority in pipeline security include enforcing its LNG safety regulations (which include specific security requirements) that existed prior to the creation of TSA. Under PHMSA regulations, LNG facilities must have, among other things, a (1) security manual, (2) security training program for employees, (3) security communications system, and (4) security lighting and monitoring system. PHMSA inspects LNG facilities to ensure they are meeting these requirements. The LNG facilities regulations are the only PHMSA pipeline regulations that specifically delineate operators' security responsibilities in detail. For hazardous liquid and gas pipelines, PHMSA has other pipeline safety regulations that require pipeline operators to prevent vandalism and unauthorized use of equipment.

While PHMSA continues to oversee the security of LNG facilities, TSA has stated that it can issue security directives, but it has not done so. These directives would allow TSA to take enforcement actions against pipeline operators. TSA currently conducts reviews of pipeline operators' compliance with voluntary guidance but does not have regulations related to pipeline security and does not take enforcement actions against pipeline operators.

To further complicate the matter, the United States Coast Guard—a DHS agency responsible for marine and port security—also has authority to oversee and enforce its security regulations for operators of LNG facilities.[8] Several of the operators' LNG facilities are located in the Nation's ports or along its eastern seaboard.

PHMSA, the U.S. Coast Guard, and the Federal Energy Regulatory Commission[9] executed an interagency agreement in 2004 for safety and security reviews of LNG facilities. The purpose of the agreement was, among other things, "to avoid duplication of effort and to maximize the exchange of relevant information related to the safety and security aspects of LNG facilities and the related marine concerns."

In our view, a similar approach should be taken with the pipeline security annex to resolve the issue of overlapping authority between PHMSA and TSA. The annex should be amended to specifically delineate the agencies' roles and responsibilities in overseeing and enforcing security regulations for LNG operators.

PHMSA and TSA Need To Maximize the Strategy Used To Assess Pipeline Operators' Security Plans and Guidance

Congress continues to emphasize the importance of securing the Nation's pipelines and related infrastructure. In August 2007, Congress passed the Implementing Recommendations of the 9/11 Commission Act. The act mandates, among other things, the following actions related to pipeline security for the Secretary of Homeland Security and the Secretary of Transportation:

- DHS, in consultation with DOT, is required to establish a program for reviewing pipeline operators' adoption of recommendations in a 2002 PHMSA security guidance document. The PHMSA guidance recommended that, among other things, pipeline operators: (1) identify critical facilities, (2) develop and implement a corporate security plan, and (3) review the corporate security plan on an annual basis and revise as necessary to reflect changing conditions. The program must also include a plan to review pipeline security plans and critical facility inspections by May 2008.
- DHS and DOT are required to develop and implement a plan for reviewing and inspecting the 100 most critical pipeline operators' pipeline security plans and

critical facilities.[10] The agencies are required to develop and implement a plan by August 3, 2008.

- DHS and DOT shall develop and transmit to pipeline operators security recommendations for natural gas and hazardous liquid pipelines and pipeline facilities by February 2009.
- If DHS determines that regulations are appropriate, it is required to consult with DOT. One of the two agencies shall then promulgate the regulations and perform necessary inspection and enforcement functions.

To determine whether additional security regulations are needed, PHMSA and TSA will need to evaluate and test the adequacy of existing security standards as agreed to under the annex. The need for new security regulations will be partly determined by the degree to which pipeline operators are following existing guidance.

However, the current security guidance created by PHMSA, TSA, and pipeline industry associations is not mandatory and remains unenforceable unless a regulation is issued to require industry compliance. Further, the guidance is general in nature and only provides a broad overview of security issues in industry and effective policies and practices. To determine if new security regulations are needed, PHMSA and TSA will need to conduct covert tests of pipeline systems' vulnerabilities to assess the current guidance as well as evaluate operators' compliance.

Although the Pipeline Inspection, Protection, Enforcement, and Safety Act of 2006 also requests an assessment of TSA security guidance, we did not perform those activities for several reasons. To truly assess whether existing security guidance is adequate, we would have needed to: (1) evaluate operators' compliance with the guidance; (2) develop testing protocols; and (3) perform vulnerability tests to ascertain, among other things, if unauthorized individuals can penetrate operators' critical infrastructure. It is not within the DOT Office of Inspector General's jurisdiction to perform such activities. We discussed these matters with the congressional committees of jurisdiction.

Because PHMSA and TSA have reviewed pipeline operators' compliance with security guidance in the past, these agencies would be best suited to evaluate this area. Therefore, it would be appropriate for PHMSA and TSA to collaborate in assessing whether existing security guidance is adequate.

In our opinion, however, actions are needed from both PHMSA and TSA to maximize the strategy for assessing pipeline operators' security plans and guidance. These actions should include the following:

- Ensuring that PHMSA is actively involved in the inspections of the 100 most critical operators' security plans and critical facilities. To date, PHMSA's role has been limited to an "as needed" basis. According to PHMSA, it has not regularly attended past TSA security reviews of pipeline operators. To ensure effective and timely execution of this congressional mandate, PHMSA should participate in these inspections on a regular basis, especially given its level of expertise in security-related matters.
- Developing testing protocols and performing vulnerability tests to ascertain, among other things, if unauthorized individuals can penetrate operators' critical infrastructure (including cyber attacks). Currently, there are no plans to develop

protocols and conduct vulnerability tests. Without testing, there is no way to effectively measure operators' compliance with existing guidance or assess the adequacy of the guidance.

PHMSA and TSA must work together to develop a pipeline security strategy that maximizes the value and efficiency of both agencies' efforts. This is a fundamental factor in enhancing pipeline security. PHMSA must make every effort possible to communicate and coordinate with TSA to move forward in completing the actions identified in this report. If PHMSA and TSA are unable to expedite the actions needed and improve coordination, it may be necessary for Congress to take action.

RECOMMENDATIONS

PHMSA should make it a top management priority to finalize and execute the action plan for implementing the security annex and meet the deadlines mandated in the 9/11 Commission Act. Therefore, we recommend that the PHMSA Administrator collaborate with the TSA Administrator to complete the following actions:

1. Finalize the action plan for implementing the annex provisions and program elements and effectively execute the action plan.
2. Amend the annex to clearly delineate the roles and responsibilities of PHMSA and TSA in overseeing and enforcing security regulations for LNG operators.
3. Maximize the strategy used to assess pipeline operators' security plans and guidance to ensure effective and timely execution of congressional mandates in the 9/11 Commission Act. At a minimum, enhancements to the strategy need to include (a) actively engaging PHMSA assistance in inspecting the 100 most critical pipeline operators' pipeline security plans and critical facilities and (b) developing protocols and conducting vulnerability tests to effectively measure operators' compliance with existing guidance or assess the adequacy of the guidance.

AGENCY COMMENTS AND OFFICE OF INSPECTOR GENERAL RESPONSE

On January 31, 2008, we provided PHMSA Headquarters officials with a preliminary draft and discussed our results with them. On February 1, 2008, we submitted our draft report to PHMSA and requested its formal comments. We subsequently met with PHMSA's Chief Safety Officer to discuss her views on the issues presented in this report. We agreed to revise the draft as appropriate to include the Agency's recent progress toward finalizing the action plan to fully implement the annex program elements. We provided PHMSA with our revisions on February 15, 2008.

We received PHMSA's formal response on March 7, 2008, which concurred with our recommendations and provided appropriate planned actions and target dates.

- **Recommendation 1:** PHMSA stated it will continue to work with its TSA partners to regularly update and prioritize items on the action plan and then execute those items. Also, to improve accountability and management visibility, the action plan will be updated monthly and forwarded to PHMSA's Chief Safety Officer for review. PHMSA's actions taken to resolve this recommendation were effective as of March 7, 2008.
- **Recommendation 2:** PHMSA stated it will work with TSA and the other agencies with which it has Memoranda of Understanding concerning LNG facilities to determine if further clarification of roles and responsibilities in oversight and enforcement of security regulations for affected LNG operators is warranted. PHMSA will also assist TSA in its communication and work with appropriate state agencies that have regulatory oversight of peakshaving[11] LNG facilities. PHMSA will notify our office of this determination by June 30, 2008.
- **Recommendation 3:** PHMSA stated that it will work closely with TSA to execute the pipeline mandates in the 9/11 Commission Act. Because PHMSA's pipeline safety inspection resources are already committed for FY 2008, it is unlikely to actively participate with TSA in the remaining inspections of the top 100 pipeline operators. However, PHMSA will work with TSA as it compiles and analyzes the results of the inspections to determine the extent to which operators voluntarily complied with 2002 security standards. PHMSA also stated that it has little experience in covertly assessing the effectiveness of security plans and must defer to TSA for this expertise. Nonetheless, based on TSA's inspections and vulnerability assessments, PHMSA will work with TSA to determine if the security standards and guidance adequately ensure the safety of the Nation's pipeline facilities and whether security-focused regulations are needed. PHMSA's actions taken to resolve this recommendation were effective as of March 7, 2008.

PHMSA also provided several general comments, such as emphasizing that, as a matter of policy, it recognizes that TSA has primary responsibility for pipeline security and it is committed to support TSA's leadership in these matters. PHMSA further emphasized recent actions taken to enhance communication and coordination between PHMSA and DHS, including the DOT Secretary and Deputy Secretary quarterly meetings with the DHS Secretary and Deputy Secretary to discuss interdepartmental issues and initiatives. PHMSA's full response is included in the appendix to this report.

We also provided TSA Headquarters officials with a copy of the draft report on February 8, 2008, and have incorporated their comments as appropriate.

Actions Required

PHMSA's response and planned actions address the intent of our recommendations. In accordance with the provisions of DOT Order 8000. 1C, we will follow up with PHMSA to ensure its corrective actions are consistent with our recommendations.

We appreciate the courtesies and cooperation of PHMSA and TSA representatives during this audit. If you have any questions concerning this report, please contact me at (202) 366-0500 or Scott Macey, Program Director, at (415) 744-0434.

Exhibit A. Objectives, Scope and Methodology, and Related Audits

Objectives

The Pipeline Inspection, Protection, Enforcement, and Safety Act of 2006 (Pub. L. No. 109-468 [2006]) directed the Office of Inspector General to review DOT's actions to implement the pipeline security annex to the memorandum of understanding between DOT and DHS. Consistent with the act, our audit objectives were to: (1) assess the adequacy and effectiveness of the process by which DOT communicates and coordinates with DHS on matters of pipeline security; (2) provide the agencies' status in implementing the program elements outlined in the annex; and (3) review the role, responsibility, and authority of DOT regarding pipeline security matters.

Scope and Methodology

We conducted the audit between June 2007 and November 2007 in accordance with generally accepted *Government Auditing Standards* prescribed by the Comptroller General of the United States. We included necessary tests to provide reasonable assurance of detecting abuse or illegal acts.

To assess DOT's process for communicating and coordinating with DHS on matters of pipeline security, we interviewed agency officials at PHMSA's Office of Pipeline Safety and TSA's Pipeline Security Division. We reviewed correspondence on pipeline security matters between the two agencies and various associated documents.

To provide the agencies' status in implementing the annex program elements, we interviewed officials at PHMSA's Office of Pipeline Safety and TSA's Pipeline Security Division to discuss actions taken and needed by the two agencies to implement the annex. We examined (1) the security annex between PHMSA and TSA to identify the roles and responsibilities between the two agencies, (2) provisions and program elements of the annex, and (3) the PHMSA/TSA action plan initiatives and target dates to implement the annex.

To determine the role, responsibility, and authority of PHMSA regarding pipeline security matters, we examined the following:

- The Aviation Transportation and Security Act of 2001 (Pub. L. No. 107-7 1 [2001].
- The Homeland Security Act of 2002 (Pub. L. No. 107-296 [2002].
- The Pipeline Inspection, Protection, Enforcement, and Safety Act of 2006.
- The Implementing Recommendations of the 9/11 Commission Act of 2007 (Pub. L. No. 110-53 [2007]).

- Title 49 Code of Federal Regulations Parts 192, 193, and 195 pertaining to pipeline security operations.
- Homeland Security Presidential Directives (HSPD) 5, "Management of Domestic Incidents," HSPD 7, "Critical Infrastructure Identification, Prioritization, and Protection," and HSPD 8, "National Preparedness."

We also interviewed agency officials at PHMSA's Office of Pipeline Safety and TSA's Pipeline Security division to obtain each agency's perspective on the division of pipeline security roles, responsibilities, and authority.

Although the Pipeline Inspection, Protection, Enforcement, and Safety Act of 2006 also requests an assessment of TSA security guidance, we did not perform those activities for several reasons. To truly assess whether existing security guidance is adequate, we would have needed to: (1) evaluate operators' compliance with the guidance; (2) develop testing protocols; and (3) perform vulnerability tests to ascertain, among other things, if unauthorized individuals can penetrate operators' critical infrastructure. It is not within DOT's Office of Inspector General's jurisdiction to perform such activities. We discussed these issues with the congressional committees of jurisdiction.

However, we did examine the 2002 Pipeline Security Information Circular, which PHMSA developed and TSA later adopted. We also examined the following:

- "Security Guidelines - Natural Gas Industry - Transmission and Distribution," developed by the Intrastate Natural Gas Association of America, the American Gas Association, and the American Public Gas Association
- "Security Guidelines for the Petroleum Industry," developed by the American Petroleum Institute.
- TSA's "Pipeline Security Smart Practices."
- Applicable laws and regulations.

We also interviewed representatives of the Interstate Natural Gas Association of America, the American Petroleum Institute, and the Association of Oil Pipeline to obtain their perspectives on the adequacy of Federal and industry pipeline security guidance.

During the audit, we visited or spoke with (via telephone) representatives from PHMSA's Eastern and Southwest regional offices to obtain their perspectives on field operations relating to pipeline security matters. We also visited five pipeline operators in Texas (Centerpoint Energy, ConocoPhillips, El Paso, Kinder Morgan, and Panhandle Energy); one pipeline operator in Virginia (Dominion); and one pipeline operator in California (Pacific Gas and Electric). At each location, we interviewed key program officials to obtain their perspectives on pipeline security matters and, where possible, toured their facilities.

Related Audit and Testimony Coverage

Report SC-2004-064, "Actions Taken and Needed for Improving Pipeline Safety," June 15, 2004. We reported that, among other things, it was unclear which agency or agencies would be responsible for pipeline security rulemaking, oversight, and enforcement and recommended that the delineation of roles and responsibilities between DOT and DHS be defined by executing a Memorandum of Understanding or Memorandum of Agreement.

Testimony CC-2004-055, "Actions Taken and Needed for Improving Pipeline Safety," June 15, 2004. The OIG presented testimony before the Senate Committee on Commerce, Science, and Transportation regarding actions taken and needed for improving pipeline safety. We reported that, among other things, DOT's Office of Pipeline Safety and TSA need to establish their respective pipeline security roles and responsibilities and then define them at the operational level.

Testimony CC-2004-061, "Actions Taken and Actions Needed To Improve Pipeline Safety," June 16, 2004. The OIG presented testimony to the House Committee on Transportation and Infrastructure, Subcommittee on Highways, Transit, and Pipelines regarding actions taken and needed for improving pipeline safety. We provided the same recommendation to the Senate Committee that we presented to the House Committee the previous day.

Testimony CC-2004-071, "Progress and Challenges in Improving Pipeline Safety," July 20, 2004. The OIG presented testimony before the House Energy and Commerce Committee, Subcommittee on Energy and Air Quality regarding progress and challenges in improving pipeline safety. We provided the same recommendation to the House Committee that we presented to the House Subcommittee on Highways, Transit, and Pipelines and the Senate Committee on Commerce, Science, and Transportation the previous month.

Testimony CC-2006-023, "Pipeline Safety: Progress and Remaining Challenges," March 16, 2006. The OIG presented testimony before the House Transportation and Infrastructure Committee, Subcommittee on Highways, Transit, and Pipelines regarding progress and remaining challenges in improving pipeline safety. We reported that the DOT's Office of Pipeline Safety and TSA still need to define their roles and responsibilities at the operational level in an annex to the MOU. A lack of clearly defined roles among the Office of Pipeline Safety and TSA at the working level could lead to duplicative or conflicting efforts, less than effective intergovernmental relationship, and—most importantly—the potential for an uncoordinated response to a terrorist attack.

Testimony CC-2006-039, "Pipeline Safety: Progress and Remaining Challenges," April 27, 2006. The OIG presented testimony before the House Committee on Energy and Commerce Subcommittee on Energy and Air Quality regarding progress and remaining challenges in improving pipeline safety. We provided the same recommendation to the House Committee that we presented to the House Subcommittee on Highways, Transit, and Pipelines the previous month.

EXHIBIT B. PROGRAM ELEMENTS IN THE PIPELINE SECURITY ANNEX

Program Element	Description
1. Identification of Critical Infrastructure/Key Resources and Risk Assessments	The agencies agreed to review existing definitions of criticality and consider the need to refine definitions. To support TSA efforts in this area, PHMSA agreed to provide compliance data, other information collected in the course of security inspections or reviews of security plans (including those required under 49 CFR 172.800), and activities of transportation carriers and shippers. Also, TSA will coordinate with PHMSA on observations or recommended measures derived from the results of criticality and vulnerability assessments, including on pipelines, to evaluate whether they conflict with or adversely affect current or planned safety requirements.
2.Strategic Planning	The agencies will seek consensus concerning measures to reduce risk and minimize consequences of emergencies involving pipeline infrastructure. Also, the agencies will identify initiatives and activities for achieving performance goals and will develop a program framework and timetable for their completion.
3.Standards, Regulations, Guidelines, and Directives	The agencies will seek early and frequent coordination in the development standards, regulations, guidelines, or directives aff-ecting transportation security; identify best practices; and explo-re opportunities to build on existing standards-setting activities. In the course of discharging their safety and security missions, the agencies will review the adequacy of existing standards in the private and public sector, identifying any gaps that should be addressed through rulemaking, guidelines, or directives.
4.Inspections and Enforcement	The agencies will explore opportunities for collaboration in inspection and enforcement activities, with the objective of maximizing the use of available resources and targeting enforcement resources on the basis of system risks. The agencies will immediately develop procedures for referral of safety and security issues to PHMSA and TSA, respectively; inventory existing inspection and enforcement resources; and develop specific plans for closer coordination in the deployment and use of inspectors, including any necessary additional training.
5.PHMSA Technical Support	TSA can ask for PHMSA's support to develop, staff, implement, or enforce regulations, orders, directives, plans, programs, or other measures. TSA can also ask for PHMSA support to conduct security reviews during an elevated security threat.
6.Sharing Information During Emergency Response	The agencies agreed to promptly share information about emergency situations that implicate the missions and interests of each other.
7.Public Communication, Education, and Outreach	The agencies will review existing protocols for public communication concerning security-related matters, specifically including review of existing protocols for publication of information contained in the national pipeline mapping system (a map of the Nation's pipelines developed by PHMSA).
8.Communicating Protective Measures to Affected Organizations	The agencies agreed to consult with one another before disseminating security requirements, voluntary standards, and guidelines that impact security to the public.
9.Research and Development	The agencies will review their safety- and security-related projects and identify opportunities to collaborate and support their strategic plan through identification, development, and testing of new or modified technologies or processes. Also, the agencies will establish protocols for ongoing information sharing and participation in their respective research and development planning processes.
10.Legislative Matters	The agencies are to consult with each other as soon as possible on the development of proposed legislation, comments on legislative proposals, draft testimony or briefings to be given before congressional bodies or staff, and answers to questions for the record.
11.Budget	The agencies agreed to communicate throughout the budget development, justification, and execution process in order to develop and present a coordinated position on transportation security funding matters and to avoid duplicative requests for funding in connection with pipeline and hazardous material transportation security.

APPENDIX. AGENCY COMMENTS

U.S. Department
of Transportation

1200 New Jersey,S.E.
Washington, D.C. 20590

**Pipeline and Hazardous
Materials Safety
Administration**

March 07, 2008

Mr. David A. Dobbs
Principal Assistant Inspector General for Auditing and Evaluation
Office of the Inspector General
U.S. Department of Transportation
1200 New Jersey Ave, SE
Washington, DC 20590

RE: OIG Project 07A30 1 8A000

Dear Mr. Dobbs:

The Pipeline and Hazardous Materials Safety Administration (PHMSA) thanks the Department's Office of Inspector General (OIG) for its efforts, reported in Project 07A30 1 8A000, to examine the complex issues of pipeline security; to interview a wide range of government and private sector stakeholders; and to render its opinions on areas of pipeline security that need improvement. This letter responds to the recommendations contained in that report.

PHMSA, as a matter of policy, recognizes that the Transportation Security Administration (TSA) has primary responsibility for pipeline security. PHMSA is committed to support TSA's leadership in these matters. The TSA/PHMSA Annex to the Department of Homeland Security (DHS), DHS/DOT Memorandum of Understanding identifies eleven Program Elements for TSA and PHMSA cooperation and collaboration to promote pipeline security.

PHMSA has acknowledged TSA's role in leading security inspections of pipeline operators and their facilities. PHMSA has, in large part, discontinued performing inspections of pipeline operators' compliance with security protocols that would duplicate TSA's inspections. We have no reports of overlapping or duplicative security inspections from pipeline operators.

Background

PHMSA's Pipeline Safety Program has always addressed the risks that security issues posed to safe operation of pipeline systems. Before 9/11, PHMSA addressed such risks by requiring operators to have plans and capabilities to protect pipeline facilities from vandalism and to facilitate response to foreseeable emergencies at their facilities.

Early in 2002, PHMSA's predecessor, the Research and Special Programs Administration (RSPA), worked with industry and State agencies, through the consensus standards process, to formulate consistent guidance for protecting Liquefied Natural Gas facilities, natural gas and hazardous liquid pipelines and their associated facilities. These guidelines identified protective actions that were scaled to the National Threat Advisory System levels.

In Section 1557 of the 9/11 Commission Act (Pub. L. 110-53), the Congress directed TSA and PHMSA to collaboratively develop and implement security standards, assess their effectiveness, then evaluate if regulations are needed to ensure pipeline security. The MOU Annex and the Pipeline Security Action Plan uniformly shows TSA as the leader in each provision, with support, collaboration, and consultation from PHMSA.

Recent Phmsa Actions

The following examples illustrate the enhanced communications and coordination between PHMSA and DHS that have been implemented in the past year. At the Departmental level, the Secretary and Deputy Secretary of Transportation meet quarterly with the Secretary and Deputy Secretary of Homeland Security to discuss cross-cutting and cross-Department issues and initiatives. PHMSA's Administrator also meets regularly with the Assistant Secretary of Homeland Security – Transportation Security Administration. Assisting TSA achieve its goals has been and remains a matter of the highest agency priority.

PHMSA representatives have met with TSA Pipeline Security Division staff at least quarterly to discuss accomplishments and milestones. In September, 2007, 9/11 Act items were placed on the Action Plan. At their meeting in December, 2007, both organizations identified primary points of contact to pass incident notification and follow-on with more detailed, operational information as events unfold. At a follow-up meeting in January 2008, changes to the Action Plan were made to reflect the agencies placing a priority on items in the 9/11 Commission Act. We plan routine coordination meetings between executives of each agency.

PHMSA recently established the new position of Planning and Preparedness Manager in the Office of the Administrator, reporting directly to the Chief Safety Officer. The manager is responsible for all aspects of PHMSA's emergency management and security policy activities and to provide liaison with other agencies, including TSA and stakeholders.. Operationally, the manager is the designated Point of Contact for incident notification to TSA's Pipeline Security Division's Policy Director.

In December, 2007, a senior PHMSA headquarters staff member accompanied TSA as it conducted a Corporate Security Review (CSR) of a pipeline operator that had experienced a leak of product caused by vandalism. The official also attended a meeting hosted by the

Commander, United States Coast Guard (USCG), DHS, Sector St. Petersburg, of the Sector's Maritime Security Council, concerning the vandalism event.

In another example, PHMSA worked with its TSA counterparts to address how the potentially sensitive information on PHMSA's National Pipeline Mapping System (NPMS), originally withdrawn from the public website after the 9/11 attacks, could be made available as an information tool for the public and community planners. As a result of this cooperation, the NPMS Public Viewer was restored on the web in April, 2007.

PHMSA participates in the DHS-sponsored Oil and Natural Gas, and Energy Government Coordinating Councils and associated Pipeline Sector Specific Council for coordination of a wide range of pipeline security issues with other Federal agency and private-sector officials. PHMSA participated in the February, 2008, DHS-sponsored Energy Critical Infrastructure Partnership Advisory Committee meeting. In December, 2007, TSA and DOT collaborated on a Transportation Security Information Sharing Plan that addresses several domains of security information, not otherwise addressed in the Annex.

Since November, 2007, PHMSA, the Federal Energy Regulatory Commission (FERC), the Department of Energy (DOE), and the Protective Security Coordination Division of the Office of Infrastructure Protection, TSA, and USCG, of DHS, have been collaboratively developing procedures to perform Comprehensive Reviews (CRs) for security of Maritime Liquefied Natural Gas (LNG) Facilities. PHMSA has existing MOUs with FERC and USCG concerning security review responsibilities at Waterfront Import / Export LNG facilities. We have discussed plans to harmonize management of inspection for these facilities. We believe that CR practices developed by this workgroup may be applied by TSA in CRs for inland, Peak-Shaving LNG facilities.

In total, PHMSA continues to communicate about pipeline incidents and facility safety information, coordinate pipeline security inspection activities, and work with TSA's Pipeline Security Division to accomplish the activities listed in the Pipeline Security Action Plan. PHMSA will improve its documentation of plans, activities and milestones and initiate regular reporting on the Action Plan to PHMSA's senior leadership. Additionally, PHMSA will continue to participate in several Government Coordinating Councils and Sector Coordinating Councils, hosted by the Department of Homeland Security's National Protection and Programs Directorate to bring its experience in pipeline safety and security to the critical infrastructure context of those Councils.

Phmsa's Response to Recommendations

Recommendation #1. Finalize the action plan for implementing the annex provisions and program elements and effectively execute the action plan.

RESPONSE: Concur: As the action plan is a dynamic document, changes and adjustments to milestones are anticipated. PHMSA will particularly pursue issues where the interplay between safety and security may give PHMSA opportunities to share its safety knowledge, operational experience, and its extensive state partnerships in support of TSA's security goals and implementation plans.

PLANNED ACTION: PHMSA will continue to work with its TSA partners to regularly update and prioritize the items on the Action Plan, and then execute those items. The last update to the Action Plan is dated February, 2008. To improve accountability and management visibility, the pipeline security action plan will be updated monthly and forwarded to PHMSA's Chief Safety Officer for review.

Recommendation #2. Amend the annex to delineate the roles and responsibilities of PHMSA and TSA in overseeing and enforcing security regulations for LNG operators.

RESPONSE: Concur. PHMSA will work with TSA and the other agencies with which PHMSA has executed Memoranda of Understanding concerning LNG facilities[12] to determine if further clarification of roles and responsibilities in oversight and enforcement of security regulations for affected LNG operators is warranted. PHMSA will notify the OIG of this determination by June 30, 2008.

PLANNED ACTION: Regardless of the determination, PHMSA will propose to incorporate LNG facility inspections into the Action Plan at the next working group meeting. PHMSA will assist TSA in its communication and work with appropriate State pipeline agencies that have regulatory oversight of peak shaving LNG facilities, which make up the majority of LNG facilities in the Nation. PHMSA also will work with TSA to determine the extent to which the data PHMSA has collected, and its state partners' field activities can inform TSA decisions on requirements of the 9/11 Commission Act and implement any resulting policy decisions.

Recommendation #3. Maximize the strategy used to assess pipeline operators' security plans and guidance to ensure effective and timely execution of congressional mandates in the 9/11 Commission Act. At a minimum, enhancements to the strategy need to include actively engaging PHMSA assistance in inspecting the 100 most critical pipeline operators' pipeline security plans and critical facilities, and developing protocols and conducting vulnerability tests to effectively measure operators' compliance with existing guidance or assess the adequacy of the guidance.

RESPONSE: Concur. PHMSA will work closely with TSA to help execute the pipeline security mandates in the 9/11 Commission Act, as listed in the current Action Plan. When participating in a TSA CSR, PHMSA inspectors participate in an advisory role, based on their safety and regulatory focus, as TSA seeks voluntary operator cooperation on security issues. However, PHMSA inspectors would address a safety issue encountered during a CSR.

PLANNED ACTION: While TSA generally advises PHMSA as they schedule CSR, PHMSA has not participated in the great majority of those reviews. PHMSA completed its FY08 pipeline safety inspection plan well before the OIG completed the investigation, and is into the second quarter of completing the scheduled inspections. PHMSA asked TSA how it intends to inspect the remainder of the top 100 operators' critical facilities. TSA stated it plans to hire consultants to assist them perform inspections of the remaining critical facilities of the top 100 operators. As PHMSA's pipeline inspection resources for FY08 are committed, it is unlikely that PHMSA would participate in these inspections.

PHMSA recognizes that the separate issues of: (1) adequacy of existing standards and guidance, and (2) voluntary compliance with those standards, are both required for successful protection of the Nation's pipeline infrastructure. As TSA and their consultants complete the inspections of the top 100 operators' critical facilities, PHMSA will work with TSA as it compiles and analyzes the results of the reviews to determine the extent to which operators have voluntarily complied with 2002 security standards.

PHMSA has little experience in covertly assessing the effectiveness of security plans that the OIG discusses in this report and must defer to TSA's Pipeline Security Division, TSA's Office of Security Assessments, and the Protective Security components in DHS for this expertise. Nonetheless, PHMSA also is very interested in working with TSA to review the macro definition used to define the top 100 operators, the variability in compliance among those operators, and the significance of any non-compliances found. PHMSA will continue to work with TSA as they determine if the 2002 security standards and guidance are adequate to ensure the security of pipeline facilities.

Based on the reviews of the Top 100 pipeline operators' compliance with existing standards, and from the review of systematic (and perhaps covert) prospective assessments of vulnerability described by the OIG in the subject report, PHMSA looks forward to assisting TSA to determine if security- focused regulations are needed to ensure the security of the Nation's pipeline infrastructure. Whether regulations are determined to be required or not, PHMSA can assist TSA use program development practices similar to those applied to PHMSA's Integrity Management Programs to help systematize our security support activities. This process is consistent with the actions detailed in the 9/11 Commission Act and also detailed in the current Action Plan.

Sincerely,

Stacey L. Gerard
Assistant Administrator / Chief Safety Officer

End Notes

[1] Pub. L. No. 109-468 (2006).

[2] OIG Report Number PT-2006-007, "DOT's FY 2006 Top Management Challenges," November 15, 2005, and OIG Report Number PT-2005-008, "DOT's FY 2005 Top Management Challenges," November 15, 2004. OIG reports are available on our website: www.oig.dot.gov.

[3] Pub. L. No. 110-53 (2007).

[4] LNG is natural gas cooled to an extremely low temperature, which causes it to liquefy. There are 113 LNG facilities in the United States.

[5] Following the events of September 11, 2001, PHMSA (formerly the Research and Special Programs Administration) developed voluntary pipeline security guidance in 2002, which TSA has since formally adopted.

[6] Liquefied Natural Gas Facilities: Federal Safety Standards, 49 C.F.R. § 193 (2007).

[7] PHMSA security guidance defines a facility's critical categorization by three factors: (1) whether it is a viable terrorist target, (2) how important it is to the Nation's energy infrastructure, and (3) how likely it is to be used as a weapon to harm people.

[8] Maritime Security: Facilities, 33 C.F.R. § 105 (2007).

[9] The Federal Energy Regulatory Commission is responsible for authorizing the site and construction of onshore LNG facilities and also conducts environmental, safety, and security reviews of LNG plants and related pipeline facilities.

[10] PHMSA security guidance defines a facility's critical categorization by three factors: (1) whether it is a viable terrorist target, (2) how important it is to the Nation's energy infrastructure, and (3) how likely it is to be used as a weapon to harm people.

[11] LNG peak-shaving facilities store surplus natural gas that is to be used to meet the requirements of peak consumption later during winter or summer.

[12] Currently, PHMSA has an MOU with the Federal Energy Regulatory Commission and the U.S. Coast Guard concerning regulatory purview of Waterfront LNG Import/Export facilities.

CHAPTER SOURCES

The following chapters were previously published:

Chapter 1 – This is an edited reformatted and augmented version of a Congressional Research Service publication, report RL33347, dated February 18, 2010.

Chapter 2 – This is an edited reformatted and augmented version of a United States Government Accountability Office publication, report GAO-10-867, dated August 2010.

Chapter 3 – These remarks were delivered as testimony given on July 14, 2010. Subcommittee on Railroads, Pipelines, and Hazardous Materials Staff, before the members of the Subcommittee on Railroads, Pipelines, and Hazardous Materials.

Chapter 4 – These remarks were delivered as testimony given on July 15, 2010. Cynthia L. Quarterman, Administrator, Pipeline and Hazardous Materials Safety Administration, before the Subcommittee on Transportation, and Hazardous Materials, Committee on Transportation and Infrastructure, United States House of Representatives.

Chapter 5 – These remarks were delivered as testimony given on June 28, 2010. Subcommittee on Railroads, Pipelines, and Hazardous Materials Staff, before the members of the Subcommittee on Railroads, Pipelines, and Hazardous Materials.

Chapter 6 – These remarks were delivered as testimony given on June 29, 2010. Cynthia L. Quarterman, Administrator, Pipeline and Hazardous Materials Safety Administration, before the Subcommittee on Railroads, Pipelines and Hazardous Materials.

Chapter 7 – These remarks were delivered as testimony given on June 29, 2010. Deborah A. P. Hersman, Chairman, national Transportation Safety Board, before the Subcommittee on Railroads, Pipelines, and Hazardous Materials.

Chapter 8 – This is an edited reformatted and augmented version of a Pipeline and Hazardous Material Safety Administration, report AV-2008-053, dated May 21, 2008.

INDEX

D

E

F

G

H

I

J

K

L

M

N

O

P

Q

R

S

T

U

V

W

Y